Granville H. Sewell

Environmental Quality Management is a perceptive and sensitive explanation of our environmental problems and the actions—regulatory, technological, and social—that can be taken to provide a stable, quality environment.

Granville H. Sewell provides a readable and knowledgeable synthesis of the biological, ecological, geographical, engineering, economic, political, and legal concepts that are required to understand man–environment interactions and to design workable programs that will reduce undesirable environmental effects while satisfying our needs for goods and services. The author focuses on major areas of environmental concern—land-use planning, water quality, coastal-zone management, air quality, solid-waste management, noise abatement, and pest control—but he also provides a methodological framework that can be applied to other environmental problems as they emerge in the future.

Environmental Quality Management

Environmental Quality Management

Granville H. Sewell

Director
Environmental Quality Management Program
Columbia University

Prentice-Hall, Inc., Englewood Cliffs, New Jersey

Library of Congress Cataloging in Publication Data

Sewell, Granville H.   1933-
    Environmental quality management.

    Includes bibliographies.
    1.   Environmental protection.   2.   Environmental
engineering.   I.   Title.
TD170.S48   363.6   75-4647
ISBN   0-13-283127-9
ISBN   0-13-283168-6   (pbk.)

© 1975 by Prentice-Hall, Inc.
Englewood Cliffs, New Jersey

10 9 8 7 6 5 4 3 2 1

Printed in the United States of America

PRENTICE-HALL INTERNATIONAL, INC., London
PRENTICE-HALL OF AUSTRALIA, PTY. LTD., Sydney
PRENTICE-HALL OF CANADA, LTD., Toronto
PRENTICE-HALL OF INDIA PRIVATE LIMITED, New Delhi
PRENTICE-HALL OF JAPAN, INC., Tokyo

CONTENTS

Rationale

This book is an introduction to the art and science of managing man's activities as they affect the natural environment. Language and, as much as possible, subjects have been chosen for comprehension by students from the social sciences, humanities, or the early stages of science or engineering education. All environmental topics have not been encompassed, but methodology has been emphasized so readers will be equipped to cope with future problems as they emerge.

Environmental management is important for modern society. A series of man-made problems in the environment--water pollution, air pollution, noise, accumulation of solid wastes, and other disquieting phenomena--has been acknowledged, and a widespread commitment has been made to control the problems. Commitment requires application of both natural and social sciences plus a spark of inspired creativity. The environmental manager needs to be able to manipulate both social institutions and appropriate technologies but must do this with the sensitivity of an artist, the insights of a poet, and, perhaps, the moral purity and determination of a religious zealot.

Organization

In our society, problem-solving implies an application of the scientific approach to specific objectives. The organization of this book reflects this background. In the first chapter, the general dimensions of the overall environmental problem are discussed. Broad social issues are identified, and terms are defined. A methodology for approaching and coping with environmental problems is explained, and the general legal and governmental apparatus devised to address the problems is described. The second chapter explains the ecological basis of our physical environment, so cause-effect relationships of man's actions can be understood and, to a limited degree, predicted.

Each succeeding chapter focuses on a specific class of environmental problems. The dimensions of the problems and their resource background are described so readers will have a sense of their priority. Then the technologies, laws, social institutions, and other resources available for providing a more stable and high-quality environment in each problem category are discussed. Material of a more technical nature or topics of specialized interest have been placed in appendices so they can be included or excluded easily by a reader. The final chapter examines two basic tools in environmental management, public information and environmental education, and then, after a brief philosophical evaluation of the book's approach, concludes with a description of opportunities available to individuals seeking a professional career as an environmental manager.

Underlying Themes

Like all books, this one possesses its individual viewpoints and biases. At least two themes are consistently woven through the ensuing discussions. One is the thesis that underlying environmental problems are primarily institutional, not technological, in nature. Citizen attitudes and expectations must eventually adjust. Governmental agencies require reorientation. Enforceable laws must be written, and implementable programs must be fashioned. Essential technology already exists, and the assumption is that improvements will follow when the resources have been mobilized and the objectives have been widely accepted. Administering this total effort is the task of the environmental manager.

A second theme expresses a profound distrust for technology as it is used in our society, which too often regards technological devices more as fetishes or symbols than tools. Automobiles tend to be expressions of status or sexual ambitions. We resemble the sorcerer's apprentice, unwittingly driven by the baubles we seek. We easily forget that machines are inanimate and only move as controlled by external forces. Yet technology provides the material foundation of our society. Without modern technology, our population would starve, freeze, and perish from innumerable diseases. So the author advocates the use of technology widely but with introspective caution.

This book also contains a moral bias towards human individuality. The underlying purpose of eliminating the smog blanket over a city is to enhance the life of the individual persons beneath and around it. We cannot examine just technical efficiencies of controls on pollution, but we must also consider how lives will be changed. We must not just calculate the gross benefits versus the costs of a specific measure, but we must ask to whom and when do these benefits and costs accrue.

There are other prejudices and themes in these pages. Even the selection and order of topics represents biases. Yet this is inevitable, especially in any exposition as profoundly political as environmental management. The important question is whether or not these themes and prejudices are adequately constructive for the society.

Use of this Book

No text is sufficient to provide a complete learning experience in an individual class. While this book is designed to provide the basic source for technical information, it is expected to be supplemented by readings and lectures that illuminate contrasting technical, moral, and philosophical positions. Graphics, particularly those that elaborate physical processes or provide examples of technologies and their effects, can be used to enliven lectures. Few experiences can substitute for the raw smells, sounds, and sights encountered by a field trip to environmental problems or facilities. And every instructor should tap the rich lode of personal experiences represented within a class of students. All of us are surrounded by an environment. The most critical problem is the development of perception and understanding. Last, class discussions that sharpen the student's consciousness and self-confidence in man's ability to influence his future should never be forsaken.

Acknowledgments

Listing all of the colleagues, friends, and students who contributed to this book is an impossibility. A few persons, however, deserve special mention, particularly Gay Darwent, Marjorie Strickland, and Betty Garbus, who each contributed to the fashioning of individual chapters. Readers whose views contributed vitally to the contents included James Listorti, Dhun Patel, and Paul Borsky. Robert Socolow, especially, took time from an incredibly crowded schedule to make numerous suggestions that removed many ambiguities and inaccuracies. Finally, special thanks must be extended to the gracious and gifted talents of Joan Giles, who prepared the masters for the pages you are reading.

A Request

No single individual can encompass the entire reaches of scholarship that can be described as "environmental studies." In this book, as in any other, certain subjects may receive less than optimum coverage and some viewpoints may appear parochial or limited. The author readily admits to these failings, but he also hopes that the readers will possess the generosity to inform him where revision efforts should be directed.

> G.H.S.
> New York City
> Winter, 1974

Environmental Quality Management

Chapter 1

APPROACHING ENVIRONMENTAL MANAGEMENT

> "We are more sensible of what
> is done against the custom than
> against Nature." Plutarch
> (c. 100 A.D.)

Basic Concepts

Meaning of Environmental Management

Environmental management is the influencing of human
activities as they affect the quality of mankind's physical
environment, especially the air, water, and terrestrial
features. The methods of environmental management can vary
immensely. "Influencing" can range from indirectly affecting behavior by altering economic incentives, such as electricity rates, to the outright prohibition of certain activities, such as the discharge of toxic chemicals into a
river. "Human activities" being influenced can vary from
the throwing of trash into a lake by a single camper or the
discharge of polluting gases by millions of motorists who
daily drive their automobiles along city streets. All active persons practice some degree of environmental management but the term is interpreted here as a conscious, systematic effort by one or more persons acting in concert to
produce an aesthetically pleasing, economically viable, and
physically healthy environment.

As a profession, environmental management has numerous
predecessors, including sanitary engineering, environmental
health, urban and regional planning, and public administration, but a separate identity has begun to coalesce in the
past several years as public agencies have expanded to im-
plement new environmental laws being passed at all governmental levels. Responsibilities are still vague. A survey
of local environmental management problems conducted by the
International City Management Association during 1973 for
the U.S. Environmental Protection Agency asked local governments to select the most applicable of four alternative
definitions of "environment." The first alternative only
considered natural aspects, such as air, noise, sewage,

1

solid wastes, toxic substances, and water. The second definition was broadened to encompass energy, historical preservation, land use, open space, radiation, population, and wildlife preservation, and the third definition added aesthetics, health, housing, mass transportation, recreation, streets, and highways. The fourth and final definition was the broadest, adding economic development, education, employment, public safety, and welfare.

While none of the four definitions was supported by more than about one-third of the respondents, a slight majority of the cities and counties favored one of the two broader definitions.* A slightly stronger concensus was found on environmental priorities. When asked by the survey to rank their environmental problems by severity, cities listed land use as the most severe problem followed by a tie between general growth problems, solid waste, and waste water for second place. Counties cited solid waste as the most severe with land use, waste water, and growth problems following.

One implication of this broad scope of environmental concern is that the managerial talent attempting to solve the problems must have equivalent breadth and an ability to synthesize the diverse technical materials. In another report entitled The Art of Managing the Environment, the Ford Foundation stated that "one of the most complex challenges" facing experiments in environmental management is the translation of technical plans into words that can be understood by the decision-making group, and a serious shortage exists for trained individuals able to cope with environmental management problems.**

The Issues

The objectives of environmental management have been largely defined by widespread public concern with a set of environmental conditions that are widely considered "problems." Some conditions are easily enumerated, such as

---

*International City Management Association, Environmental Management and Local Government, a report to the U.S. Environmental Protection Agency (1974).

**Ford Foundation, The Art of Managing the Environment, a Ford Foundation Report (September, 1974).

visible smog banks in the cities, reports of health effects
from pollutants in air and water, visual litter, noisy
streets, descriptions of vast oil spills and their effects,
clogged highways, and disappearing rural landscapes. Other
conditions are more vague, including speculations about the
effects of possible climatic changes or dire material shortages.

To cope with these conditions, including those currently unidentified but presumed on the basis of past experience
to exist or potentially exist, we have identified various
forces or factors that are considered to be causative.
These have become the foci of environmental management efforts. There are several levels of causation. Attention
is usually preoccupied with immediate factors, such as
spreading suburban sprawl, industrialization with attendant
pollution, and the inadequacy of various social institutions, such as laws and enforcement agencies, to maintain
a quality environment.

Underlying these factors, however, are the public
values and attitudes that have produced and perpetuated
undesirable environmental conditions. Environmental protection requires conscious effort by individuals living and
working within the environment. Resources must be diverted
from other uses to provide protection. Consumer products
must be redesigned and, in some cases, not produced. Habits
must be changed. Expectations and standards of living must
be adjusted, although the adjustment may be eventually
viewed as more desirable. Each of these subjects will be
addressed in more detail within the following pages.

The Essential Glossary

Significance of Words

Words are tools that we use with varying motives and
different degrees of precision. Most words used in discussions on environmental topics should be viewed with a
healthy degree of skepticism and tolerance. "Environment"
and "pollution," for example, are imprecise. Their meaning
will vary from person to person and occasion to occasion.
Also, both have emotional connotations. Compare "pollution"
with "contamination," or "environment" with "surroundings."
"Pollution" has a nasty tone, a touch of the obscene, while
"contamination" is more neutral. "Environment" evokes a

3

more romantic imagery than "surroundings." These words are firmly rooted in the public's lexicon, and their total meanings should be considered if communication--a key element in widespread environmental improvement--is to occur.

Environment

In dictionary terms, environment can be defined as the sum of all external influences and forces acting upon an object, usually assumed to be a living being. For the word to be useful, an object must be identified before the environment can be analyzed. For mankind in general, the word would encompass the air, water, land, vegetation, various animals, and any other matter, force, or influence within the planet or outside that could affect a person's life. Should we include the moon? The moon causes tidal action in the oceans and may be involved in other geological or biological--or even psychological--phenomena we have not yet identified. And the definition could be extended to the solar system since nearly all activity on the earth is ultimately dependent on the sun's rays. In other words, the environmental limits cannot be strictly defined, although the object within the environment usually can. "Environment" for most analytical purposes, however, refers to the biosphere, the zone of the earth's surface, waters, and atmosphere inhabited by living organisms.

Environment can have numerous dimensions. There is a social environment, the relationships between individual organisms within a single community. We can also refer to an environment evolving through time, the life environment of a particular person or society. "Environment" sometimes has operational meanings. In environmental health studies, environment is described as the intermediary zone of influencing factors that lie between the agent, which causes an accident or disease, and the host, which suffers the mishap. Thus, an automobile accident can be prevented by shaping an environment that will cause the drivers to act in a different manner, or a disease can be prevented by environmental intervention--such as sanitation--before a pathogenic (disease-causing) organism can reach a host.

Pollution

Several schools of thought have attempted to define pollution. Most controversy has revolved about the degree to which mankind should be the focus of the definition. The narrower view particularly associated with engineering fields considers pollution to be any waste discharges or even natural environmental changes that are directly detrimental to man. For instance, several cans and bottles cast beside a highway would be pollution, but the same cans and bottles placed in the middle of a desert never visited by man would not be pollution.

Another view is that we are incapable of defining what will be detrimental or not. Insects that are disease vectors may be harbored by the rubbish in the desert, and certainly the objects will affect the types and behavior of wildlife in that immediate vicinity. In any case, we cannot predict where man's future activities will lead him, so pollution can also be defined as ". . . an undesirable change in the physical, chemical, or biological characteristics of our air, land, water, that may or will hostilely affect human life or that of other desirable species or industrial processes, living conditions, and cultural assets; or that may or will waste or deteriorate our natural resources."*

A third and growing view rejects the egocentric emphasis on man and his desires. Thus, pollution can further be defined as any disruption by man of the natural system. Ethically, the disruption should be stopped or minimized regardless of the effects on man. Yet this definition demands an immense philosophical adjustment for a culture that views mankind as privileged among all living beings. Also, it must be translated into policy cautiously because, just by existing, man is disrupting the nonhuman environment, which is synonymous in many minds with natural environment.

Possibly the broadest definition for pollution is "something out of place." By not specifying the alien material or criteria for "out of place," it covers a range of interpretations. Natural pollution, such as gases from a volcano, are included. The narrow margin between a valuable something and a polluting something is recognized. For example, materials discarded by man typically have an economic value if they could exist at an appropriate place in a suitable form. Cans and bottles are valuable if massed at

---

*National Academy of Sciences/National Research Council, Waste Management and Control, Publication 1400 (1966), page 3.

a recycling plant or if filled. They are usually considered worthless otherwise. Physically, the metal and glass are virtually unchanged. Flyash is a pollutant in the air but a resource when added to concrete as aggregate. Dust swept from cultivated land clogs machinery and smothers vegetation, shifting from a soil resource to a form of pollutant.

Measurements of Pollution

Some forms of pollution are obvious. Trash littering a highway is unquestionably "out of place." But the more typical forms of water and air pollution are less easily detected until their environmental impact has already been registered. Identifying and measuring pollutants remains among the most persistent problems in environmental improvement. Many pollutants, such as trace metals or chlorinated hydrocarbon insecticides, must be measured in parts per billion, which means that extreme care is needed in analysis and even slight contamination of analytical glassware, instruments, or chemicals will distort data. Sometimes several pollutants will be acting synergistically, the total effect being greater than the sum of individual pollutants acting independently, so identification of one pollutant-- or even several--may still be insufficient to explain a disruption in the environment. Furthermore, the amount of a pollutant may differ immensely from one point in an environment to another a short distance away. Just a few inches in water may mean different conditions because of currents, temperature, and similar influences. In summary, all precise numbers reported to indicate amounts of a pollutant in an environment should be viewed with suspicion until all aspects of collection and measurement are thoroughly understood.

Another difficulty in measuring or predicting levels of pollution is the appearance of secondary pollutants, products of a chemical reaction between a pollutant and constituents of the environment. The secondary pollutants of automobile combustion, for example, are the acids, ozone, and other materials formed when gasoline fumes and other exhaust gases rise into the atmosphere and combine under sunlight with the nitrogen and oxygen. They cannot be measured by simply monitoring the original pollutants. Similarly, some pesticides can be degraded in the soil to chemicals that are more toxic than the original.

Methodology of Environmental Management

The Art and Science of Problem Solving

Functionally, the environmental manager solves--or, preferably, prevents by prior planning--problems of an environmental nature. Environmental management is partially science and partially art. Scientific principles are usually involved in the cause-effect relationships that account for a problem's appearance and solution. Computer simulation can be a powerful tool in predicting behavior of natural bodies, such as rivers; and econometric (mathematical economics) methods provide valuable insights into costs and benefits. A "scientific approach" is used for problem-solving in general.

Environmental management problems, however, inherently involve people. Implementation of decisions must be conducted by people. Environmental decisions are intrinsically political, determining winners and losers for scarce resources, so a political type of involvement by people and their communities is inevitable. Yet human values interpreting winning and losing are ill-defined and frequently contradictory both within an individual and between individuals. Circumstances, including those created by environmental management, will influence the individual responses.

In addition, our knowledge of environmental situations is never complete. We always lack sufficient data for predicting future events, and guessing what one does not know is part of the art. When the abstractness of environmental quality characteristics--and the conflicts that arise with other social priorities--are also considered, the constant need for creativity in environmental management is obvious. The approach to each situation must be tailored to the individual physical, social, political, and economic circumstances.

Despite the distinctiveness of each problem, though, our society has worked out a systematic means of approaching problems. Steps can be formulated in different ways, but about five stages are usually listed. This does not mean that the steps are always sequential or unrelated. As part of the artistic aspect, environmental management is a process in which each step must be anticipated while actions and responses are continually monitored and actions adjusted.

7

1. Define the Problem

All dimensions of the current or anticipated situation are studied. For this, information must be collected. Nothing substitutes for total immersion in a problem situation as a means of identifying the full range of considerations. If the problem consists of community noise, data describing the pattern (time, level, objectionable features, location, and any other possibly associated factors) should be collected and analyzed, preferably by someone who will later be able to interpret the data and recall further information that may not have been recorded.

Gradually, the objectives or goals--the description of the desired situation--must be defined at least in generalities, and this may require further research. For example, tentative noise-level objectives may be set by balancing known psychological and physiological effects and implications of various noise levels against the estimated technological, economic, and political costs of attaining those levels. These initial or hypothetical objectives can and should be adjusted as the process progresses and more information is obtained, but an initial estimate is essential to provide guidance on the information that should be obtained.

2. Identify and Analyze Possible Actions

Based on suspected cause-effect relationships, numerous actions could be taken for any problem. Some actions will have short-term effects; others, long-term effects. Some will have exhorbitant costs; others, low costs and, probably, slight effects. From the numerous possibilities, one selects those that appear generally acceptable on a total benefit-versus-cost basis and studies them in detail to determine which--if any--should be implemented.

Systematic identification of possible actions is vital since too narrow a conception of choices is a frequent and disastrous error in perpetuating weak solutions to environmental problems. A time sequence; e.g., short-term, middle-term, and long-term solutions, provide one type of framework for systematically identifying possible actions. In many cases, a more morally defensible and comprehensive technique is to move from the offending technology to the innocent, affected party. Most undesirable environmental phenomena today have a technological source, such as noise-making machinery or air-polluting industrial processes. Beginning

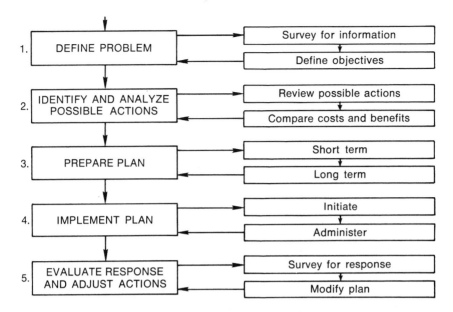

Figure 1-1. The problem-solving approach
in environmental management.

with the justification for retaining the technology, one can
ask a series of questions:

(1) <u>Removal</u>. Can the technology be simply removed and
not replaced without suffering significant loss? Example:
Removal of noisy and redundant equipment.

(2) <u>Replacement</u>. Can the technology be replaced by
another, less objectionable form to perform the same function? Example: Replacement of automobile by mass transit.

(3) <u>Relocation</u>. Can the technology be moved to an area
where it would be less objectionable? Example: Relocation
of a highway away from a community.

(4) <u>Modification</u>. Can the technology be modified or
redesigned to make it less objectionable? Example: Replacement of a standard automobile engine with a stratified combustion engine to reduce air pollution.

(5) <u>Supplementation</u>. Can another technology be added
to reduce the objectionable aspects? Example: Adding a
muffler to an internal combustion engine.

(6) Enclosure of Source. Can the technology be enclosed to intercept the pollution? Example: Covering a
noisy generator with a soundproof building.

(7) Erect Interceptors. Can some type of barrier be
constructed to prevent passage of the objectionable pollution between the offending technology and the receiver?
Example: Planting trees and constructing walls along noisy
streets to intercept noise.

(8) Shielding the Receiver. Can the receiver be physically enclosed to prevent the objectionable aspects of the
technology from reaching him? Example: Soundproofing residential buildings.

(9) Personal Protection. Can the individual utilize
personal protection devices, such as earmuffs or face masks?
Example: Provide ear muffs to airport technicians.

Note that the first few alternatives place the cost and
responsibility for correcting the situation upon the individuals responsible for the pollution, not upon the innocent
receiver. Gradually the costs are shifted toward the receiver and innocent community. These are not exclusive
alternatives. Some could be undertaken simultaneously, but
most would be in a different time frame. Many of these
alternatives would obviously not be feasible, but the purpose of the sequence is to provide a system for comprehensive review, not the most desirable solutions.

Desirability of a solution has to be judged by a modified cost-benefit analysis in which all the costs--quantita-
tive and qualitative--that can be reasonably anticipated to
result from a particular decision are compared to all the
benefits calculated similarly. Ideally, a standard cost-
benefit analysis comparing a series of possible actions will
enable a planner to maximize the present value of all benefits minus costs. "Present value" means that anticipated
future benefits and costs are discounted (diminished by some
interest rate for each year the project is expected to
exist). The rationale is that the same resources placed
into a bank account or other dependable investment with compound interest would in, for example, twenty years produce a
predictable amount of additional resources that could then
be used for social purposes. To justify foregoing the benefits of this safe investment, the environmental project's
costs and benefits should be discounted at the same interest
rate.

To avoid double counting, only primary benefits (those directly attributable to the project) and not secondary benefits (those that result from the primary benefits) are used. For example, the added income of sporting goods stores around a new state park are secondary benefits already counted in the value of park visitations. However, the nature and scale of secondary effects, such as the probable prospering of sporting goods stores or the impact on local law enforcement agencies, should be noted as part of the overall analysis. See the references in Suggested Readings at the end of this chapter for more detailed instructions on performing cost-benefit analyses.

In environmental planning, a formal cost-benefit analysis can only serve as a crude guideline and even then a decision-maker should critically examine each calculation of benefits and costs. At least five questions should be asked.

(1) If the analysis culminates in a ratio of benefits to costs or in simple dollar figures, what basis has been used for quantification? Oversimplified or incomplete indicators often have to be used. To account for loss of human life, for example, the economic loss of salaries and wages during the person's anticipated life time has customarily been used. Value of housewives has often been estimated by the salary that would have to be paid to a maid to perform the same housekeeping tasks. Thus, a human life becomes equated with the cost of bulldozers or other items of property.

(2) How are intangibles considered? Aesthetics in general have been notoriously difficult to include in a standard cost-benefit analysis. Psychological confidence is an integral part of the aesthetic experience. Water can be purified and treated until it is chemically indistinguishable from wholesome, natural water, but the psychological factor of knowing the water is treated will still diminish enjoyment for most drinkers. Similarly, the depth of aesthetic experience by an individual to a rare environmental feature, such as a landscape or the spring freshness of forest air, and the relationship of this experience to the well-being of both the individual and the larger society has never been satisfactorily reduced to a quantitative scale.

(3) What discount or interest rate is used? Most costs normally occur in the beginning of a project when property is purchased and facilities, such as roads, are constructed, but benefits begin later and occur over many years.

Therefore, a low discount rate will favor the project but a high rate will act against it. A related question is the estimated life span that will be assigned to a project. In the engineering field, a life of twenty to fifty years is customarily used for major projects, such as dams or highways. Some major environmental projects, such as a park, can last indefinitely and even then escalate in real-estate value.

(4) How are risks considered? The future cannot be predicted, and every major project is normally plagued by unexpected mishaps. These normally raise costs. There is an old adage that every disaster is accompanied by a hidden opportunity. Once a project has been initiated, however, the implementing organization already possesses a set of commitments that usually limits its ability to seek and use possible opportunities. Several analytical techniques are used to anticipate these events. A sensitivity analysis, for instance, is the changing of one or more of the assumptions in the analysis to test the effects on calculated costs and benefits. Another method is to calculate an "envelope," an estimate of the benefits and costs under both the worst and best reasonable sets of circumstances. The most probable estimate of events presumably falls somewhere in the middle.

(5) Who receives the benefits and who pays the costs? Answers are sometimes relatively clear. Costs of pollution control equipment installed by an industry will presumably be paid initially by the industry, and all individuals who were suffering damages, either in property loss or health, will initially benefit. In other cases, such as the creation of parklands, the answer becomes more complex. Millions of persons may benefit directly, but this benefit will vary immensely. For some visitors, it will be slight and partially transferable to other parklands. The few persons evicted from the parkland will often pay an intensely emotional cost that cannot be calculated in the fair-market compensation.

In practice, cost-benefit analyses serve functions in addition to providing guidance for decision making. For example, they are usually designed to establish credibility among those readers or listeners whose cooperation is required for implementation of any decision. In any case, the alternatives must be sufficiently understood in all their implications that they can be woven into an overall action plan.

3. Prepare Plan

The next stage in problem solving is to arrange the possible actions into a plan or strategy that will represent the most effective use of resources for attaining the desired objectives. For example, an examination of total costs and benefits in a problem involving street noises may indicate that altering the noise-making characteristics of trucks could provide the highest degree of immediate relief from the noise, but a combination of eliminating some truck use and rerouting remaining trucks would be a desirable long-term program.

4. Implement Plan

Implementation demands a set of management skills often not needed in previous stages of problem solving. Commitment must be translated into action; and orchestration of personnel rather than perception, creativity, and persuasion with ideas becomes the primary requisite. General management principles apply. See Suggested Readings for examples of general management textbooks.

5. Evaluate Response and Adjust Actions

Every program encounters unanticipated changes in circumstances. Social objectives shift, physical conditions prove other than anticipated, and even the impact of successful action necessitates either a change in objectives or a termination of the effort. As industries and towns, for example, construct pollution-control facilities, the work of the local pollution-control agencies must shift from pressuring for construction to the policing of effectiveness from existing facilities.

Discontinuities in Planning

Inherent Problems

The tasks of environmental management are complicated by numerous inherent methodological problems. Several have already been discussed, but at least three others, (1) geographical jurisdictions, (2) categorical divisions, and

(3) use of standards, are also normally encountered in an environmental program.

Geographic Jurisdiction

Geographically, legal jurisdictions rarely coincide with an environmental problem. In theory, they will probably never coincide, because interests in local use of resources--including land--are divided between successively larger and, to some extent, competing communities. And, since environmental systems can never be closed from the rest of the world, a local problem typically has effects far beyond local boundaries. Thus, environmental management must include efforts to persuade other authorities to cooperate and to coordinate activities.

Pollution Problems Tend to be Interrelated

Solid wastes can be burned, creating an air-pollution problem. Or solid wastes dumped in a kitchen sink can become water pollutants--until they settle to the bottom of a tank or stream and become a solid-waste problem again. Yet the laws and environmental protection agencies are usually divided along categorical lines: air, water, noise, radiation, solid-wastes, and similar problem areas. And this characteristic appears likely to continue because environmental problems are manifested in categorical ways, such as piles of waste, pollution-laden air, or visible fish kills in a stream.

Use of Standards

Quantitative standards, the maximum or minimum acceptable limit set for most environmental parameters, are subject to change and frequently do change. And the direction of change is almost invariably towards stricter levels, causing dissatisfaction among persons who have complied with earlier standards. Yet this tendency is innate in the nature of standards, as explained in the appendix to this chapter. They are set by examining criteria, the empirical, descriptive evidence of effects that different levels of a pollutant may have upon the environment, including human health. Then--in theory--the standard is arbitrarily set by specialists who, by studying criteria, weigh the total known threat of the pollutant against the consequences of

14

stricter standards. In practice, many standards have
evolved through time and are not significantly changed by
specialists because adequate criteria are lacking and disruptive political forces would be unleashed.

Environmental Laws at the Federal Level

Recent Legislation

. Over the past decade, an elaborate structure of federal
legislation has been enacted; and many of these laws have
been adapted by the state legislatures for state use. A
partial list would include the Air Quality Act of 1967, the
Clean Air Act Amendment of 1970, the National Materials
Quality Act of 1970, the Solid Wastes Disposal Act of 1970,
the Occupational Safety and Health Act of 1970, the Water
Pollution Control Act of 1972, major revisions in the Insecticide, Fungicide, and Rodenticide Act, the Coastal Zone
Management Act of 1972, the Marine Protection Research and
Sanctuaries Act of 1972, and the Noise Control Act of 1972.
In addition, other legislation was passed focusing on related fields, including outdoor recreation, wildlife conservation, and transportation. Most of these acts will be discussed in later chapters, but one, the National Environmental
Policy Act of 1969, has had a particularly wide impact upon
governmental policy decisions and, more than any other legislation, is demanding the attention of professional talent at
the policy-making level.

The National Environmental Policy
Act of 1969 (NEPA)

NEPA has been both blessed as the most effective device
ever enacted to force consideration of environmental factors
within executive decision making and soundly cursed as a
diabolically destructive impediment to efficient governmental administration. The target of these opinions is the
requirement in Section 102 (C) that all agencies of the
federal government

> include in every recommendation or report on
> proposals for legislation and other major Federal
> actions significantly affecting the quality of
> the human environment, a detailed statement by
> the responsible official on (i) the environmental

impact of the proposed action, (ii) any adverse
environmental effects which cannot be avoided
should the proposal be implemented, (iii) alternatives to the proposed action, (iv) the relationship between local, short-term uses of man's environment and the maintenance and enhancement of
long-term productivity, and (v) any irreversible
and irretrievable commitment of resources which
would be involved in the proposed action should
it be implemented.

The act contained two major sections, or titles, the
first declaring a national environmental policy and the
other establishing the Council on Environmental Quality.
Briefly paraphrasing the policy section, the act states the
general national objective to be obtaining productive harmony between man and nature and fulfilling needs of both
present and future generations. Six specific objectives
are listed: (a) responsibility to future generations, (b)
provision of a quality environment for all Americans, (c)
prevention of undesirable impacts, (d) preservation of national heritage, (e) achievement of a population-resource
balance, and (f) enhancement of renewable resources and recycling of nonrenewable ones. Each person, according to the
act, should both enjoy a healthful environment and contribute to maintaining it.

Besides the impact statements, federal agencies are
required to (a) use systematic, interdisciplinary approaches
in planning and decision making affecting the environment,
(b) develop procedures for considering unquantified environmental values, (c) avoid unresolved conflicts over alternative uses of land, water, or air, (d) cooperate internation-
ally for maintaining environmental quality, and (e) bring
all operating regulations into compliance with the act. But
the clear, action-forcing mechanism of the impact statement
is lacking except for Section 102 (C).

In the first three years of NEPA's existence, over
3,000 environmental impact statements were filed with the
Council on Environmental Quality. The quality of most was
relatively poor, but the statements were available for public scrutiny; and they had to be circulated to other governmental agencies for comment before becoming effective.
Thus, a form of quality control did exist. Three strategies
typically reduced the effectiveness of the statements.

Black-Box Approach. The necessity to consider alternatives was interpreted to be akin to the cost-benefit analysis that had long guided investment by the Corps of Engineers and similar organizations. Thus, an effort was made
to quantify all parameters of environmental quality and, as
far as possible, reduce them to a single number so one project could be compared to another. Intangible considerations suffered. Furthermore, while citizen groups could
sometimes unravel the rationale and find appropriate information on how a decision was made, this often took considerable technical expertise and placed questioners at a disadvantage.

Limited Assumption Approach. In considering alternatives, possible alternatives that would remove a project
from an agency's jurisdiction were ignored. For example,
a highway department would consider as alternatives the
routes A, B, and C. Alternative modes of transportation,
such as rail, or drastically different approaches to the
problem would not be identified.

Primary Impact Approach. Many impact statements ignored
the secondary effects that a project would cause by stimulating economic growth in an area or by subtle disturbances
of interrelated ecosystems. A highway department, for example, would consider erosion, wildlife disturbance, visual
changes, and similar direct intrusions of the highway. But
often ignored were the secondary effects of industrial and
residential development that would be attracted along the
highway's path, especially at interchanges with local roads.

It should be emphasized that the effect of NEPA is limited to actions of the federal government. State actions
are not affected unless the federal government is partici-
pating in the specific project. Private investments are not
included unless federal permission or financing is involved.
Many states and local governments, though, are trying to
extend the concept to their jurisdictions.

The Existing Environmental
Management Structure

Overview

The decisions affecting environmental quality are currently made within a political process that is largely

dominated by actions of three organized groups--government, industry, and consumer or special-interest organizations. Governmental agencies at the federal, state, and local levels have regulatory responsibility over all actions significantly affecting the environment. Because of the provision in the U.S. Constitution that states are automatically vested with all powers not specifically reserved for the federal government, their power could be considered theoretically dominant. In practice, though, the federal level has preempted authority in numerous areas neglected by state action, and the management standards continue to be generally set by federal agencies.

Federal

At the federal level, the Environmental Protection Agency was established in December, 1970, to consolidate into one agency the major federal programs dealing with air pollution, water pollution, solid waste disposal, pesticides regulation, and environmental radiation. Omitted were management of public lands, operations of the U.S. Corps of Engineers, some environmental health programs, administration of the Environmental Education Act, occupational health programs, and a number of similar, more specialized activities.

The Environmental Protection Agency reports directly to the President, being independent of any regular department. Also reporting directly to the President is the Council on Environmental Quality, a small, policy-level group established by the National Environmental Policy Act of 1970 with the purpose of reviewing all environmental activities and making policy recommendations to the executive branch. While the legislative and judicial branches have developed considerable expertise in environmental questions, these are still largely handled by the previously existing structure.

State

Many states have followed the example of the federal government and have created environmental protection agencies in the executive branch. In a few cases, these are known by other names--such as departments of conservation-- or the responsibilities are distributed through other branches, such as departments of health and departments of natural resources. In a study conducted during 1974 by the

Conference of State Sanitary Engineers, air pollution control was found to be a responsibility of environmental protection agencies in twenty-two states, health departments in fifteen states, natural resource departments of six states, and under an independent commission in seven states. Water pollution was within environmental protection agencies in twenty-two states, health departments in thirteen states, natural resource departments in eight states, and independent commissions in eight states. Solid waste management was divided almost equally between the state health departments and nonhealth departments, while noise abatement was primarily under state health departments.

Local

Reorganization in recent years has occurred less at the level of local governments. Solid wastes are typically under the jurisdiction of departments of public works or departments of sanitation. Air pollution sometimes has its own department; and, in other localities, it is under the department of health. Water supply and sewage treatment may be under health, public works, or a separate agency. Often a water or sewage district is formed to provide a regional service with separate administrative authority. Land-use planning, if it occurs, is usually performed by a local planning board.

Industry

Virtually all the major corporations with production facilities have established environmental departments. Authority granted to these departments by the policy-making management varies, but recognition has spread that environmental concern is a serious corporate responsibility. In large corporations, change has been accelerated by the Occupational Safety and Health Act of 1970, which emphasized the need to maintain a healthy work environment.

Citizen Groups

Public-spirited environmental groups, such as the Sierra Club, Friends of the Earth, the National Audubon Society, and numerous other organizations with a general, special-interest, or legal-advocacy orientation, have spearheaded the pressure for changing both the governmental structure and public attitudes towards the environment. Despite shortages of funds, frequent organizational problems, and occasional losses in public credibility, they remain the most effective assurance that the movement towards environmental improvement will continue.

Environmental Management in Perspective

Maintaining environmental quality is important for society and becoming more important. But, for persons involved in environmental programs, a slightly detached perspective mixed with considerable patience and humor may be needed. Other social objectives are also vital--health, housing, social equity, education, and nutrition. Furthermore, political support for environmental action is constantly shifting. Environmental improvement has become a vague symbol of achievement for many groups with narrowly defined goals, such as improvement of camping conditions in national parks, but uncertain reliability on other issues. And the ranks of environmental advocates are crowded with many who simply want to express their multiple frustrations with our technological society.

Perhaps, too, a degree of hypocrisy exists in our pressures for a cleaner environment. Our society has reached the highest living standards in human history. We expect clothes to be whiter than white, foods to be perfection in appearance, and transportation to have speeds and reliability that would tax gods. These achievements are fashioned by wringing resources from the environment with preciously few resources allotted to covering up our damage. Yet we do not want the environment to show the effects.

A cynic viewing our society could ask whether the development effort is worth the cost. Compared to some of the few "primitive" societies remaining on our earth, we have not fared exceptionally well in our search for satisfaction or that elusive quality, happiness. Lives of contemporary society seem crowded with stresses, insecurity, and frustrations. Yet there are a few cultures still existing with annual monetary incomes of a few dollars per capita--or less--and they seem cheerful, sleep soundly at night, cannot visualize violence or crime, and appear at ease in their physical environment. We are inescapable prisoners in our cultural situation, but it is worth realizing that a clean environment will not alone produce a euphoric utopia.

QUESTIONS FOR DISCUSSION

1. Describe how personalities that we have elevated to folk heroes reflect specific social values that can be extended to our treatment of the environment.

2. If you were to plan a future city to be an environmental utopia, to what physical features and social institutions would you give particular attention? Explain your reasoning.

3. Explain the statement, "Pollution is an institutional problem." (A dictionary defines institution as (a) an organization or establishment, or (b) a pattern of group behavior.)

4. What is the local governmental structure regulating water pollution, air pollution, solid waste disposal, and noise abatement in your home town?

5. If studies suggested convincingly that 50,000 deaths in the U.S. annually now attributed to emphysema, bronchitis, malignancies, and pneumonia could be prevented by halving levels of air pollution, would you vigorously advocate the necessary changes in life style and public investments? Would you do the same if the report indicated that deaths attributable to automobile accidents could be halved from 50,000 to 25,000 annually by installing engine governors that would limit all automobiles to forty miles per hour?

SUGGESTED READINGS

The literature that may interest a reader concerned with environmental management in its broadest scope can be divided into about four categories. The first category--like this volume--uses a survey approach to identify environmental problems and explain common technological solutions. Each author or editor has an individual approach. In the following two examples, the orientation is suggested by the title. Environmental Health has an emphasis on disease prevention while Environmental Engineering and Sanitation provides more technical details. Urban Environmental Management draws more material from the earth sciences. All are recommended as references.

1. P. Walter Purdom (ed.), Environmental Health (New York: Academic Press, 1971).

2. Joseph A. Salvato, Jr., Environmental Engineering and Sanitation, 2nd edition (New York: Wiley-Interscience, 1972).

3. Brian J. L. Berry and F. E. Horton (eds.), Urban Environmental Management: Planning for Pollution Control (Englewood Cliffs, N.J.: Prentice-Hall, Inc., 1974).

The second category of readings expounds on the contribution that can be provided by individual disciplines, such as law, economics, psychology, and medicine, towards the solution of environmental problems. Examples would be:

4. Robert Dorfman and N. Dorfman (eds.), Economics of the Environment: Selected Readings (New York: W. W. Norton and Co., 1972). Paperback. Excellent quality and range of environmental economics readings.

5. A. R. Prest and R. Turvey, "Cost-Benefit Analysis: A Survey," The Economic Journal (December, 1965), pp. 685-729. A classic review of weaknesses inherent in cost-benefit analyses.

6. H. G. Thuesen, W. J. Fabrycky, and G. J. Thuesen, Engineering Economy, 4th edition (Englewood Cliffs, N.J.: Prentice-Hall, Inc., 1971). Detailed explanations for the mathematics of cost-benefit analyses.

7. Abram S. Benenson (ed.), Control of Communicable Diseases in Man, 11th edition (Washington, D.C.: American Public Health Association, 1970). Paperback. A standard reference compendium covering 117 diseases in over 296 pages. Dry, cryptic, but recommended for all professionals who are responsible for health aspects of environmental management.

8. Albert Mehrabian and James A. Russell, An Approach to Environmental Psychology (Cambridge: M.I.T. Press, 1974). Relatively technical textbook for exploring relationships between sensation and environment.

9. Frank P. Grad, Environmental Law: Sources and Problems (New York: Mathew Bender, 1971). The lawyer's law book on environmental law.

10. Arnold W. Reitze, Jr., <u>Environmental Law</u> (Washington, D.C.: North American International, 1972). An expensive but helpful review of environmental law. Is suitable for the nonlawyer.

In organizing and administering an environmental agency, one of the standard references on organizational management may prove useful. Examples would include:

11. R. A. Johnson, F. E. Kast, and J. F. Rosenzweig, <u>The Theory and Management of Systems</u>, 2nd edition (New York: McGraw Hill Book Co., 1967).

12. Douglas McGregor, <u>The Professional Manager</u> (New York: McGraw Hill Book Co., 1967).

The third category of readings uses case studies or examples to describe how the disciplines can be synthesized. Two competently prepared studies are:

13. Elizabeth Haskell and V. S. Price, <u>State Environmental Management: Case Studies of Nine States</u> (New York: Praeger Publishers, 1973).

14. Richard A. Cooley and G. W. Smith (eds.), <u>Congress and the Environment</u> (Seattle: University of Washington Press, 1970).

The fourth category comprises the more philosophically inclined treatises that try to understand the genesis of the environmental problems. Two drastically different examples are:

15. Rene duBois, <u>Man Adapting</u> (New Haven: Yale University Press, 1965).

16. Barry Commoner, <u>The Closing Circle</u> (New York: A. Knopf, 1971).

The Two Sides of Environmental
Quality Standards

Blessing or Curse?

Standards--those published lists of permissible parts
per million or micrograms per cubic meter of a pollutant in
our air, water, food, or other material--are alternatively
demanded and then condemned by representatives of our society. Standards represent easily understood and clearly
defined goals for environmental purity. In a court of law,
they are enforceable. They have the shorthand function of
informing us quickly whether a particular environment is
probably acceptable.

Yet standards have severe limitations. They are typically defined by experts who have pored over meager, inconclusive, and scarcely appropriate information. Standards
are used to protect a population that varies drastically in
susceptibility to injury. Standards often represent a mixture of scientific judgment, economic and political sensitivities, and aesthetic and moral taste. From a viewpoint
of simple logic, standards have at least three obvious
flaws: (1) artificial isolation of a pollutant, (2) limited
time considerations, and (3) applicability to a limited
population. When an attempt is made to apply standards to
an environment by ordering in effect, "Thou shalt not pollute this environment more than x level," we are plunging
ourselves into a practical, moral, and legal dilemma.

1. Isolation of Pollutants

A single pollutant is never found alone. It exists
with other compounds in varying quantities with all condi-
tions--temperature, movement, biochemical reactions, and
similar parameters--changing at a multiplicity of rates.
If we try to measure the effect of the pollutant in the
real world, we never can be certain that an unrecognized
"hidden" factor and not the pollutant is causing the damage.
If we use an oversimplified world, such as a laboratory with
human or animal subjects, we face the problem of extrapolating from this unreal situation to the vastly more complicated conditions in human communities. Usually, too, we

cannot work with data from humans. Ethics will not permit us to deliberately poison humans to identify the lethal dosages. But the physiology of animals is different from humans, and we cannot be certain that humans will react similarly.

2. The Time Factor

Ideally, the experts establishing a standard will use--either explicitly or implicitly--a dose-response curve (see Figure 1-2) for a specific population of humans (or other environmental biota) exposed to larger and larger quantities of the concerned pollutant. As the quantity "q" (dose) increases, the symptom "Y," which is usually the most serious effect reported in the literature, is measured. The time factor is only considered marginally. Most information in the literature, though, is collected from clinical sources, which means that people were observed only while in a laboratory, factory, hospital, or otherwise within the range of observation of the researcher. These studies are usually necessarily limited in time. Perhaps another more serious effect would have emerged if the subject had been given a lower dose and the researcher had waited longer for the effect to become apparent. Cancer-causing agents, for example, may not be operative for several decades. Genetic effects, too, may not become identifiable for one, two, or more generations.

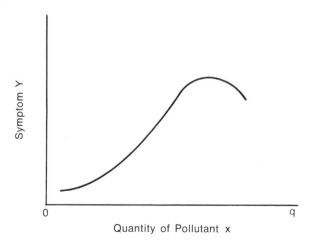

Figure 1-2. A typical idealized dose-response curve for the effects (y) of a pollutant (x) on biota.

3. Limited Population

    Each individual has a different sensitivity to a toxic pollutant, such as sulphur dioxide. When a population is exposed to a moderate air-pollution episode, the effects will be found to differ for each person. A few may become severely ill or die, but a few others may exhibit no ill effects even when studied carefully. The majority of the people will probably fall between these two extremes. If severity of effect were plotted on a graph, the line can be expected to have a bell or "normal" shape. The individuals at the extreme end of "severe effects" would be the elderly and the chronically ill with respiratory diseases. On the "no effects" end, naturally resistant individuals and young persons would be found (see Figure 1-3).

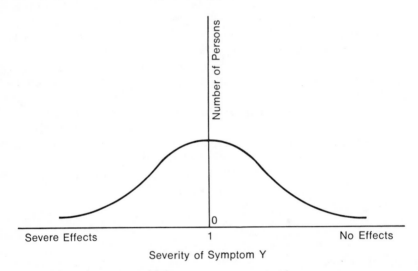

Figure 1-3.   Typical, idealized distribution of a population suffering symptom Y from exposure to a toxic pollutant.

    When a standard is set for a common technology, such as the automobile, the standard cannot realistically be defined as zero. People will insist upon using the technology regardless of someone being harmed. But any value greater than zero for the standard will mean that some part of the population will be harmed. It follows that this vulnerable population and their advocates will constantly press in political channels for stricter standards. Thus, standards in our society have rarely been fixed but have tended to

gradually become stricter though never completely satisfying
the critics.

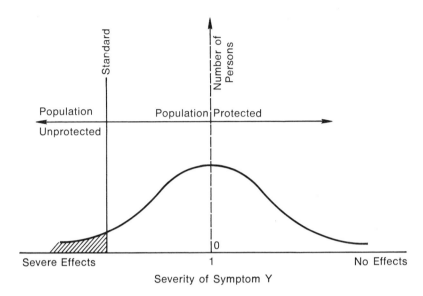

Figure 1-4. Effects of population protection when
an environmental standard is set at
any level above zero effects for any
individual in population.

Ambient Standards

Ambient standards, the levels to be found in the natural environment, were favored by both the air and water quality legislation of the sixties. Several problems arose,
some practical and others theoretical. A pollutant is never
distributed evenly over an environment. It tends to concentrate in pockets, but these are always shifting in space and
time. Obtaining measurements that will withstand legal attacks and be reproducible by other scientists is a difficult
task.

On a more theoretical level, an ambient standard does
not automatically identify who is responsible for its maintenance. An incredibly complex situation can evolve. For
example, assume that the ambiance standard for a specific
pollutant has been set at Q_0 parts per million, which is
significantly above the levels currently found in a river.
Years pass and industrial growth continues. The ambient
level of the pollutant approaches Q_0. Now a new factory

27

will be built, but they cannot be permitted to discharge any of the pollutant because the ambient standard would be exceeded. Construction of a factory with zero pollution would be considerably more expensive than experienced by the other factories, placing the new factory at a severe disadvantage. Nor can the older factories be expected to voluntarily modify their plants to permit the newcomer a share in the pollution. Anger would mount; and, sensing an impossible situation, the enforcement authority would probably overlook the transgression of the new factory if they polluted a reasonably short reach of the river.

Standards in Perspective

These weaknesses have often been noted. William T. Sedgwick, an early public-health authority, referred to standards as "devices to keep the lazy mind from thinking." He observed that "Standards are often the guess of one worker, easily seized upon, quoted and requoted, until they assume the semblance of authority."[*] Abel Wolman, a legendary figure in environmental health, stated at Congressional hearings in 1960, "Administrators are constantly searching for mathematical certainty in the solution of many of these complex problems, despite the fact that there is a high degree of uncertainty in the underlying scientific principles that are involved in establishing such standards. They are characterized by moving empiric decisions, by persistent reappraisals, by consistently giving the public the benefit of the doubt, by an ever narrowing gap between knowledge and application, by qualitative rather than quantitative slide rule assessments of hazard, and by objectives calling for the elimination of fatalities due to disease."[**] As C. P. McCord commented, "It may be recognized that we have made a

---

[*] M. Levine, "Facts and Fancies of Bacterial Indices in Standards for Water and Food," Food Technology, 15:29 (November, 1961).

[**] U.S. Congress, 2d Sess., "Radiation Protection Criteria and Standards," hearings before the Special Subcommittee on Radiation of the Joint Committee on Atomic Energy (May–June, 1960).

fetish of exact standards under circumstances where exactness is not in order."[*]

Despite the weaknesses, standards will continue to be used as expressions of objectives for environmental quality. They identify goals we intend to achieve or maintain in suppressing a particular pollutant. They will be misunderstood, and frustrations will be expressed by politicians and public alike. Industrialists and other potential polluters will attack them from one side, while the victims of pollution will be challenging them on the other. Yet even this open debate about the quality of the environment has positive effects, and the process of setting a standard can be a vital stage in attaining environmental quality under a democratic political system.

---

[*]C. P. McCord, "Man and His Environment--A Conference Resumé," Industrial Medicine, 30:354 (September, 1961).

Chapter 2

ECOLOGY AND ENVIRONMENTAL CHANGE

> "Ah, how unjust to Nature and
> himself is thoughtless, thank-
> less, inconsistent man!"
> Edward Young (1742)

Basic Concepts

Definitions of Ecology

In environmental management, ecology represents both a science and a mode of thinking. Ecology can be defined as the study of interrelationships between living organisms and their environments. The word is derived from the Greek oikos, "house" or "dwelling place," and the suffix for study, "-ology." Some authorities have referred to ecology as the study of the structure and function of nature or the science of the living environment. The German biologist Ernst Haeckel in 1870 described it as knowledge concerning the economy of nature. The range of concern is profoundly broad and, until recently, was relatively removed from practical issues.

This breadth of thinking provides a framework for considering implications of possible actions affecting the environment. Every action has a reaction. In the study of ecology, nature is envisioned as a vast web of interdependent life, each species adapting to the complex climatic, geographic, and biological environment in a manner that will ensure survival. Each has overlapping, sometimes competing needs. And the distribution of each is controlled by the existence of a unique set of environmental conditions, including temperature, moisture, energy sources, and competition, known as its niche. A rule in ecology is that no two organisms can occupy precisely the same niche in the same locality without one being eventually eliminated by the competition.

While nature is normally considered to be in a state of equilibrium, gradual change does occur without human intervention. By human scale, this change can be slow, such as the geological circulation of minerals through rock cycles; or change can be spectacularly swift. Mount Tambora in Indonesia erupted in 1815, darkening the skies with an estimated 150 cubic kilometers of ash. New England had snow storms in June and frosts in July and August. Crops failed and Europe suffered record-breaking cold, upsetting the natural rhythm for countless species of plants and animals. A sudden mutation (spontaneous genetic change) of a common pest, such as the mosquito, that would provide protection against human controls could prove troublesome--or even disastrous--to the human population.

Usually, however, the rate of natural change in the environment is sufficiently slow that it can be disregarded in human planning, especially when compared to change being precipitated by man's activity. Removal of trees from a steep hillside can cause swift erosion, a shift in the surviving plant species, a different pattern of terrestrial wildlife, and changes in the stream life where silt is deposited. Under some circumstances, even climate can be affected if forests are depleted, and more profound changes then follow.

Because of the widespread need to predict and control these changes, the popular meaning of ecology has shifted perceptively in recent years. For almost a century, ecological studies focused primarily on the taxonomy (classification) of nature's structure and the narrow functions of specific species. Distribution and diet of a bird species compared to other species, for example, would have been a typical topic of study; and textbooks emphasized the importance of classifying "natural" relationships between species of biota and the nonliving environment. The field was considered a basic, descriptive science, not a predictive tool for policy purposes. And, since most ecologists had turned to ecology in preference to man-oriented studies, they were often not inclined by temperament or training to venture predictions about consequences of human actions. In their support, the extreme complexity and paucity of information about environmental relationships will always make predictions risky, at best. Applied ecology, the art of predicting environmental reactions, has gradually spread, but the strains within the professional field of ecology are still evident.

Elements of the Ecosystem

Indicative of this shift in ecology has been the increasing use of the term ecosystems, a group of organisms
and aspects of their environment that function coherently
together as a system. Of course, no global ecosystem is
self-sufficient since, as a minimum, all depend eventually
upon the sun as an energy source.

A human body is, in a sense, an ecosystem because it
involves a series of units--protoplasm, cells, tissues,
organs, and complementary organisms, such as intestinal
bacteria--operating in a coherent fashion with a minimum
input and output. But more commonly used examples of ecosystems are (1) a pond with its water, minerals, plant life,
fish, and small animals living in a long-established relationship with each other, or (2) a forest in which trees
nourish insects, birds, earthworms, bacteria, and fungi
while these, in turn, furnish services and materials necessary for the trees.

All ecosystems can be divided into two broad sets of
components, the biotic (living) and the abiotic (nonliving).
Examples of the abiotic parts would be the water, air,
minerals, and various inert gases. The biotic category can
be subdivided into three functional groups:

(1) Producers. Green plants, algae, phytoplankton (the
smallest unit of plant life) are known as autotrophic (self-
nourishing) organisms that form the basic building blocks of
an ecosystem. These are really converters or transducers
that assemble essential minerals, capture the sun's energy--
usually by photosynthesis--and build the complex organic
compounds necessary for all living tissues, thus transforming light energy into chemical energy. By far the greatest
volume and variety of producers can be found in the seas,
but the earth also harbors an immense variety, including the
trees, grasses, shrubs, and man's food crops.

The scale and significance of the producers' role is
suggested by an analysis of the initial stages in the energy
cycle. Almost all energy originates from the thermonuclear
reaction within the sun. As hydrogen is transmuted into
helium at the sun's surface, radiation--energy moving at or
near the speed of light--scatters into the solar system.
About half of these energy pulses can be described as visible light. The earth receives about 1/50,000,000th of the
sun's energy. Roughly half of the energy pulses striking

the atmosphere eventually reach the earth's surface. Of the
energy lost, about two-thirds is reflected from clouds, 20%
from dust, and another roughly 10% is absorbed by ozone,
oxygen, water vapor, and other atmospheric compounds or is
scattered by air molecules.

Only about 1% or less of the sun's energy striking the
earth's atmosphere is actually converted by producers into
plant tissue. While this may seem an insignificant amount,
it is sufficient to produce 100 bushels of corn per acre on
an Iowa farm, not including the plants, birds, fungi, bacteria, and insects that depend upon the corn production.
The immensity of this achievement is suggested by calculations indicating that about 20,000 pounds of carbon dioxide
must be extracted from 21,000 tons of air to produce the
100 bushels of corn.

(2) Consumers. All animal life, including insects,
mammals, fish, birds, and man, ultimately depend upon consumption of producers to sustain life. They are the hetero-
trophic (other-feeding) organisms. Consumers can be further
subdivided into (a) primary consumers (herbivores) that
utilize plant life directly--cows, deer, grasshoppers, and
many fish; (b) the secondary consumers (carnivores) that
obtain their nourishment by feeding on the primary consum-
ers--cats, dogs, and carnivorous species of fish; and (c)
the mixed consumers (omnivores) that do not discriminate
sharply in their food choices between producers and other
consumers, a category including man, many birds, and some
fish.

(3) Decomposers. Essential in almost any ecosystem are
the decomposer organisms, the fungi and the bacteria that
break down complex compounds from waste materials--including
dead producers and consumers--to make the chemical components again available to producers. Often producers and
consumers will cooperate with decomposers. For example, a
fallen tree may be attacked by boring beetles. Bacteria,
water, and fungi will then enter the beetles' tunnels and
other small insects will become embedded along the protective tunnel walls.

As the loosened pulp is softened by water, possibly
aided with freezing, bacteria can begin penetrating more
deeply. Meanwhile, beetles and moisture will be loosening
the bark so that snails, slugs, and various beetle larvae
can begin stripping the log's protective covering. As wood
is softened, boring beetles can no longer survive and are

replaced by fungus-eating beetles or various larvae that burrow only in softened wood. Gradually the molds and bacteria following the paths of insects reduce the wood to a soft mound of humus that is broken apart by the roots of plants and is gradually distributed across the forest floor by water, earthworms, ants, and various other insects or their larvae. This process of one specialized group of organisms in an ecosystem being succeeded by another group is an example of succession.

Example of an Ecosystem

A small pond on a farmer's field can provide a familiar example of an operating ecosystem (Figure 2-1). The abiotic

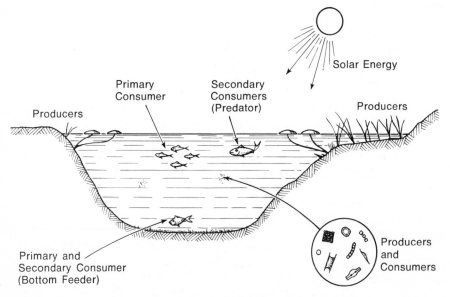

Figure 2-1. Like a home aquarium, a pond provides a convenient example of an easily defined ecosystem.

materials are the water and the carbon dioxide, oxygen, calcium, nitrogen and phosphorus salts, amino humic acids dissolved in the water, the rock and earth substrate, the ceiling of air, and a complex flow of trace elements and compounds. Of the more than ninety elements found in nature, thirty to forty are required by almost all living organisms. In agriculture, for instance, success or failure of some crops under suitable climatic conditions can often be attributed to availability of a trace mineral, such as iron or zinc. Within the pond, only a small portion of these abiotic materials are available to life. Most materials are trapped in the mud as insoluble compounds, and the rate of release for the materials can be a key to the rate that life develops within the water.

Producers. Along the margins of the pond will be emergent vegetation, the rooted plants with leaves that protrude above the water. As the depth of water increases, the type of plant gradually changes. Rooted plants eventually have floating leaves and, finally, as the depth becomes too great for plants to maintain long stems, all vegetation becomes nonrooted. Most are species of algae. Typical examples would be the various types of microscopic diatoms with silica shells, green algae with one or more cells, and--if temperatures, acidity, and nutrient conditions are favorable--a tremendous mass of simple, often single-celled blue-green algae. The latter become particularly important in domestic pollution because they can convert gaseous nitrogen into nitrates and, when other necessary nutrients are available, can literally fertilize the pond.

This plant life in the pond varies widely in appearance, size, and numbers, but all types share a common functional characteristic. Each is using the process of photosynthesis to combine carbon dioxide and other substances into complex organic compounds while releasing oxygen. Where light cannot penetrate and photosynthesis is impossible, living vegetation will not be found. If the waters are turbid (cloudy), plant life will be limited to the upper layers. And since photosynthesis only occurs during daylight, oxygen will not be released at night--although plants will continue to respire, consuming the valuable and limited oxygen.

Consumers. A variety of consumers also exists in the pond. Among the animal groups represented will be the mollusks, zooplankton (free-floating animals), and nekton (swimming organisms)--including crustacea, insect larvae, and the fish. If our pond is in the south, there will be a significantly larger number of frogs, water snakes, turtles, and other larger biota. Each species occupies a relatively precise functional niche within the pond's ecosystem. Foods may differ; habitats may be distinct; and preferences for temperature, oxygen levels, and protection from predators may contrast. Smaller fish and insects will usually depend upon phytoplankton and zooplankton for nourishment, while larger fish will prey upon the smaller fish. Bass, perch, and, especially, trout--if any exist--will seek higher levels of oxygen while carp and catfish will be less dis-criminating. Depending upon their ecological requirements, some fish will retreat to the lower, cooler strata of water during the day while others will remain near the surface. Some will rarely venture from the bordering emergent vegeta-tion while others will swim freely in the open water.

Decomposers. Last, decomposers will be present both in the water and the mud. Some bacteria will be aerobic (oxygen-dependent) bacteria while others will be anaerobic (independent of oxygen). Decaying fish, fecal matter, and plants will be broken apart and their constituent chemicals returned to the waters.

The Dynamic Balance. While life in the pond may be in equilibrium, it will never be static. Plant and animal life will be emerging, growing, and dying at rates critically influenced by the surrounding environment. Rainstorms will wash organic materials and minerals into the pond. Floods may scour the bottom and mix the waters. Sunny days will favor some species, while cool days will encourage others.

Over twenty-four-hour periods, temperature and oxygen levels will vary. Superimposed upon these daily cycles will be the seasonal variations of summer, fall, winter, and spring. And further superimposed will be the long-term geological cycles as the sediments forming the pond's mud gradually mount, deep-water life diminishes and disappears, and the pond is gradually transformed into a marsh and, eventually, dry land. Then a new geological age or cata-clysmic event may create a pond or sea again.

Ecosystem Dynamics

In an ecosystem, no component endures unchanged. All elements are in some phase of a cycle. Nevertheless, cer-tain cycles are more critical in the short-term period or are more vulnerable to man's interference. In terms of human concern, the most important are probably: (1) the hydrologic cycle, the biogeochemical cycles of (2) carbon, (3) nitrogen, and (4) phosphorus, and (5) the energy flow or food web.

(1) Hydrologic Cycle. The hydrologic cycle is the move-ment of water molecules from the earth's surface into the

atmosphere by evaporation and their return by rain, snow, or
other form of precipitation. Although the fundamentally up-
and-down cycle may seem simple, details become complex.
Evaporation from the sea and other water surfaces depends
upon the air's temperature, movement, and existing humidity
(moisture content). Plants contribute to the air's moisture
by transpiration, and discharge of water vapor through walls
of exterior cells. Between 97 and 99% of the water entering
the plants from the soil evaporates; and, for every pound of
dry material produced by the plant, up to 1,000 pounds of
water is transpired. Most agricultural crops use about 500
pounds or more of water for each pound of dry matter,
although drought-resistant crops may use as little as 250
pounds. Desert plants still transpire but become dormant
when water is absent.

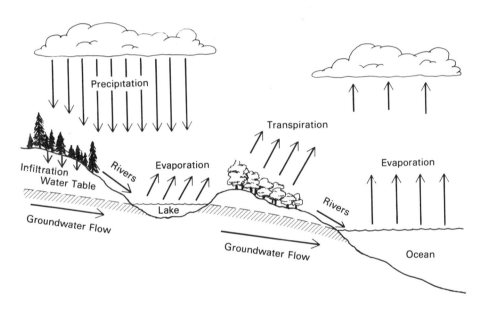

Figure 2-2. The hydrologic cycle. (From McCabe and Mines,
Man and Environment, Vol. I, Prentice-Hall, Inc., 1973.)

The return of water to the earth also is governed by a
series of environmental conditions. Prevailing winds, especially when they become fickle, are notorious for blessing
some regions with rain while depriving others. Where a
range of coastal mountains shields the interior from winds,
the seaward slopes are typically soaked with water while--as
in the state of Washington--deserts exist inland. As winds
force water-laden clouds into the cooler atmosphere of the
mountain peaks, the moisture condenses and rain falls,

leaving the winds moistureless for the land of the "rain-shadow" beyond. The troublesome drought of northeast United States during the mid-sixties was due to a shift in the dominant winds, particularly at higher altitudes.

When rain or snow strikes the ground, some is absorbed by the soil and eventually goes into the water table or is taken up by plants. The remainder represents run-off, joining the springs and seepage from the water table to form lakes and rivers.

(2) Carbon Cycle. Life in any ecosystem is largely determined by the type, number, and growth of producer organisms. In turn, the development of producers can be related to availability of the ingredients and conditions necessary for protoplasm, the complex organic structure of protein and other organic and inorganic substances forming the material of the living cell. Besides light energy and appropriate temperatures, formation of protoplasm requires a series of elements known as nutrients. Carbon, hydrogen, oxygen, nitrogen, potassium, and phosphate are typically most influential on performance of the ecosystem. Of these, the carbon atom represents the basic building block for organic molecules. Note that three of the nutrients--carbon, hydrogen, and oxygen--are readily available in a usable

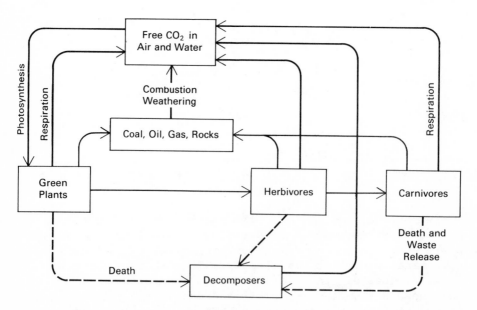

Figure 2-3. The carbon cycle. (From McCabe and Mines, Man and Environment, Vol. I, Prentice-Hall, Inc., 1973.)

form within air and water, but the others are not. Thus, commercial fertilizers typically have three primary ingredients--nitrogen, phosphate, and potash.

The carbon cycle is especially elegant since it involves both a solid and a gaseous stage and is about balanced between return and removal from the biosphere. As carbon dioxide, carbon is drawn from the atmosphere during photosynthesis by producers, passes through consumers and decomposers, and then usually reenters the atmosphere. The atmosphere, though, is only about 0.03 to 0.04% carbon dioxide and depends upon the oceans, which may store more than 50 times as much as the atmosphere, as a reservoir. Another significant reservoir, which is not available to producers, is the carbon locked in rock--such as limestone--and fossil fuels, including peat, coal, and petroleum. As these are burned, they theoretically increase the level of carbon dioxide in the atmosphere, but the function of the earth's waters in maintaining an acceptable equilibrium between the constituents of the air is not fully understood. For instance, while absorbing carbon dioxide for photosynthesis, some aquatic plants are depositing carbon as calcium carbonate, which eventually becomes mud and rock. For example, 200 pounds of Elodea canadensis is able to generate as much as four pounds of calcium carbonate in ten hours of sunlight. But the full extent of this mechanism is not understood. Meanwhile, weathering, burning, and volcanic activity are decomposing limestone and returning some carbon to the atmosphere as carbon dioxide or carbonic acid.

(3) Nitrogen Cycle. Like carbon, the nitrogen cycle has both solid and gaseous phases, but these similarities are deceptive. The behavior and functions of the compounds are different. Unlike carbon, nitrogen is the most common atom (79%) in the atmosphere. However, only a few simple, specialized organisms--certain bacteria and blue-green algae--can use this atmospheric gas directly. Most plants must have a supply of nitrogen "fixed" into more complex compounds, particularly nitrates but also nitrites, ammonia, urea, protein, and nucleic acids. While some fixation of nitrogen occurs through lightning, sunlight, and other chemical processes, as much as twenty times the fixation is caused by biological mechanisms. Bacteria in the root-nodules of some plants are among the most common and best known. But other bacteria, fungi, and algae are also important sources. The role of blue-green algae apparently depends upon complex environmental conditions.

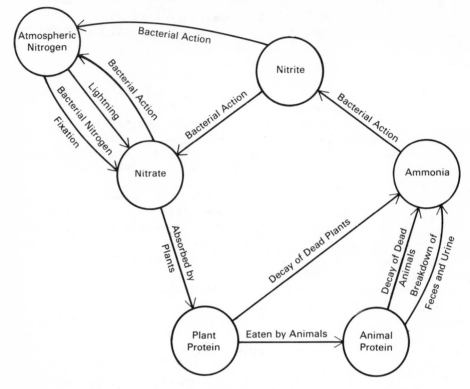

Figure 2-4. The nitrogen cycle. (From McCabe and Mines,
Man and Environment, Vol. I, Prentice-Hall, Inc., 1973.)

While biological processes can make nitrogen available
to an ecosystem, other organisms--the denitrifying bacteria--
can be working against the process. Also, nitrates are
highly soluble and are constantly being washed from the soil
and deposited in the ocean as sediments.

(4) Phosphorus Cycle. Of all the essential elements,
phosphorus is typically the one in least supply. As phosphate, the element moves rapidly through available stages in
the environment, quickly becoming locked in sediments or
biological forms--such as teeth or bones--that are not
quickly available again. The ratio of phosphorus to nitrogen in natural waters is about one to twenty-three. In
agriculture, the primary sources are the phosphate rocks
that have been deposited during ocean sedimentation of previous geological eras and the guano deposits, the dung of
seabirds. Within nature, organic phosphate from living or
dead organisms is decomposed into an inorganic form, which
can then be incorporated into new protoplasm.

(5) <u>The Food Chain</u>. All life in an ecosystem is linked to producers through a food chain, the sequence of consumption that passes energy and vital compounds from one organism to another. Many food chains occurring within a single ecosystem form the <u>food web</u>, the complex network of possible consumption choices. In each food chain, a plant must initially form energy-containing compounds and assemble minerals. A primary consumer eating the plant will reject certain substances, such as cellulose, and will incorporate other compounds into its own tissues. A secondary consumer eating the primary one will again juggle the components, incorporating some and rejecting others.

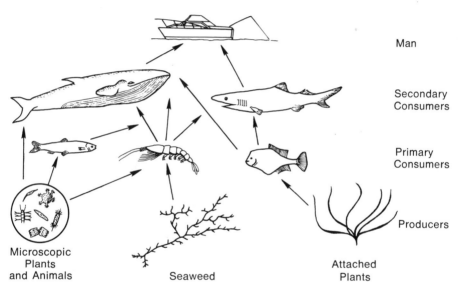

Man

Secondary
Consumers

Primary
Consumers

Producers

Microscopic
Plants
and Animals

Seaweed

Attached
Plants

Figure 2-5. Example of a marine food web.

What happens if the animal's digestive organs erroneously select the wrong materials? Over extended periods, the organism adjusts or perishes if the inappropriate material accumulates in tissues. Radioactive strontium, for example, is not easily distinguished by digestive systems from calcium, and the radioactive substance may be mistakenly concentrated in bones and teeth. DDT, by contrast, becomes magnified because it is absorbed by fat tissues. If the tissues are consumed by a predator, the DDT is absorbed by the new host where it becomes increasingly concentrated, a process known as <u>biomagnification</u>.

41

Each time the stored energy passes from one organism to another, most--usually on the order of 90%--is lost in the transfer. This constantly diminishing amount of energy available to successive trophic levels in the food chain creates the energy pyramid, the dependence of any species on the continued existence of a vastly larger pool of energy at the previous trophic level. Omnivores, such as man, have the advantage of being able to choose food sources from several different trophic levels. For example, a cornfield that can support about ten men a day when the corn is first fed to cattle can support 100 or more men when they eat the grain directly. But ecological efficiency should not be confused with nutritional adequacy, which a purely corn diet would not have.

Limiting Factors

The principle of tolerance states that an organism can survive only when its limits--both maximum and minimum--for all essential environmental factors are not exceeded. Sometimes restricted substitution is possible. A mollusk can survive a shortage of calcium if another mineral, such as strontium, is available as a partial substitute. Of course, the same organism may be able to tolerate wide variations in one aspect of the environment but not in another. A human can live without food energy for longer periods than without water--and for considerably longer than without oxygen. Nor can generalizations be made for degrees of toleration between organisms for the same factor. A camel is obviously more tolerant to water shortages than a human. Depending upon the environment, though, generalizations can frequently be assumed for the probable limiting factor. Oxygen is rarely limited in terrestrial environments, but easily becomes critical in water.

The concept of minimum requirements was first reported in 1840 from systematic studies by Justus Liebig, who discovered that plant growth was often limited not by nutrients but by micronutrients--the many elements that plants require in trace quantities. Later researchers expanded the principle to include oxygen, temperature, and even time. V. E. Shelford after 1911 added the consideration of maximum tolerance.

From these observations, the concept has gradually evolved that stability in an ecosystem is related to its biological diversity. A wide diversity, each organism having a distinct set of limits, implies a broad spectrum of tolerance for environmental change by the biota as a whole and a higher probability that the system will survive disasters. If, for example, a sudden spring freeze devastates one species of vegetation, diversity improves the probability that other species will survive to feed consumers, prevent erosion, and hold together components of the ecosystem.

Man's Interference in the Global Ecosystem

All organisms modify their ecosystems to some degree. But man has placed stresses of unprecedented variety and scope upon his environment. Some stresses are almost unavoidable in maintaining today's culture. In agriculture, uniculture--one crop concentrated across many acres--is practiced for efficiency in feeding urban dwellers. But diseases and predators can sweep unhindered through the fields, flocks, and herds. Land clearing and even selective timbering have reduced the variety of habitats and, thus, the possible diversities in wildlife.

Another type of stress has been the interference in natural cycles through uncontrolled discharge of industrial wastes. Because of the amounts of fossil fuel that we burn, for example, some observers have suggested that we may be overloading the atmosphere with carbon dioxide. At the same time, pollution of the earth's water surfaces, where possibly 70% of the photosynthesis and carbon dioxide reduction occurs, may be compounding the problem. If carbon dioxide levels in the earth's atmosphere were to rise, temperatures should also increase since carbon dioxide allows direct sunlight to pass but absorbs the reflected heat from the earth's surface. Instead of warming, though, there is evidence that the earth may be cooling, possibly because of greater dust levels in the atmosphere due to agricultural practices, widening desert zones, and increased use of the atmosphere's upper stratas by aircraft. Yet even these cause-and-effect relationships can be questioned since sunspot activity and other natural solar cycles may still be dwarfing man-made influences.

In other cases, pollution simply exceeds the limits of tolerance of some biota. For example, power plants and factories discharging heated water into rivers can destroy some of the aquatic biota, particularly trout and a few other species sensitive to heat, and alter the river's ecosystem. Unfortunately, many implications of pollution may still be

beyond our understanding. For example, rainfall records at
La Porte, Indiana, indicate a rise of about 30% in precipitation since the 1930s. Nearly 40% more thunderstorms,
usually in the early morning, are being reported. Presumably these changes can be attributed to dust from the steel
mills of Gary, Indiana, about 30 miles away, the dust providing nuclei for formation of raindrops. But unknown is
the impact of change's in rainfall and cloud cover upon local
flora and fauna.

Several examples of the dangers in industrial-type pollution can be cited. One involves the lead levels that are
known to be rising in the urban atmosphere. This could be
affecting weather, and it certainly affects health in some
instances. Some observers have waggishly suggested that our
famed devotion to the automobile and its internal combustion
engine, that still usually emits lead as a pollutant except
from recent models, may be leading us along the path of the
Roman Empire's demise. Then wine from lead-lined containers
was synonymous with the Good Life because the flavor was
distinctly smoother and more satisfying than that of wine
taken by peasants from common crockery. Only in this century have we realized that the wine dissolved minute quantities of lead and a mild form of lead poisoning probably
contributed to the sterility and ineffectiveness of the
upper classes during the Empire's decline.

Most of the ecological horror stories, though, have
involved food chains, especially with the intrusion of pesticides. A legendary case was publicized by Rachel Carson in
her complacency-shattering book, Silent Spring, in 1962.
Clear Lake in the mountains of California about ninety miles
north of San Francisco was favored by fishermen and vaca-
tioners. Despite the name, the lake was cloudy because of
a soft, black sediment covering its shallow bottom. Among
the various organisms nurtured by this mud was the larvae of
a small midge that, while not known to bite people, could be
an exasperating nuisance for anyone trying to share its
habitat.

In the late forties, someone proposed using the newly
developed chlorinated hydrocarbon insecticides to reduce the
midge population. DDD, an insecticide considered less dangerous than DDT, was carefully added to the lake in an
amount that would not represent more than one part of the
chemical to each seventy million parts of the water. Within
two weeks, no DDD could be detected in the lake waters and
the midge population almost disappeared for several years.

44

In 1954, the treatment was repeated with a slightly stronger dosage, one part of insecticide per fifty million parts of water.

That winter, the western grebes--diving birds somewhat resembling a duck--along the lake's shores began to die. Bodies of more than 100 were reported. More grebe deaths followed a third pesticide treatment of the lake in 1957. When the fatty tissues of the grebes were then analyzed, biomagnification of the pesticide was found to have occurred, and analysts found about 1,600 parts of the pesticide per million parts of tissue. During the winter months when food was scarce and the birds resorted to using their fatty tissue for survival, the pesticides destroyed cells in a vital organ and the bird died. Nor is this story unique. Other examples have shown that the concentration of a chlorinated hydrocarbon may be magnified a million times above that of surrounding waters by the tissues of predatory birds.

The Open Ecosystems

The astute environmentalist draws a series of conclusions from a review of ecology. As stated earlier, every action in the environment produces reactions. Everything can be assumed related to everything. Closed systems in the sense that we can identify a single result of our actions do not exist. The food we eat, the fuels we burn, and the other resources we harvest consist of materials that exist within the environment before, during, and after their use. But reconciling this realization with the practical realities governing our lives requires an intensive examination of specific problem topics.

QUESTIONS FOR DISCUSSION

1. Describe an ecosystem that you know, identifying (a) materials that are recycled within the ecosystem and (b) materials that typically flow into or out of the system. Does man have an identifiable role?

2. Sketch the (a) carbon or (b) nitrogen biogeochemical cycle, and identify points where man may be interfering.

3. Using the ecological concept of limiting factors, describe at least four methods that you could--in theory--

safely use to eliminate roaches from a kitchen. (Chemical biocides are not to be used.)

4. In ecological terms, describe what happens when raw sewage is released into a typical nonpolluted stream or pond, if the sewage has (a) turbidity (cloudiness), (b) a massive population of active decomposers, and (c) nutrients.

5. Ecologically, what differences would you expect in the effects between pollution in the air and pollution in the water?

6. Identify environmental costs associated with use of (a) "tin" cans for containing vegetables, (b) disposable paper towels, (c) nylon shirts, and (d) the automobile for suburb-city commuting.

SUGGESTED READINGS

An annual summary of the environmental situation in the United States is prepared by the Council on Environmental Quality and is available from the U.S. Government Printing Office at a modest cost. For example, in 1974 the Environmental Quality--1973 could be ordered as Stock Number 4111-0020 from the Superintendent of Documents, U.S. Government Printing Office, Washington, D.C. 20402, for $4.30.

Ecology textbooks or references that would assist the environmental manager could include:

1. Edward J. Kormondy, Concepts of Ecology (Englewood Cliffs, N.J.: Prentice-Hall, Inc., 1969). A readable, reasonably complete yet inexpensive paperback in ecology.

2. Eugene P. Odum, Fundamentals of Ecology, 3rd edition (Philadelphia: W. B. Saunders, 1971). The classic ecology textbook with an emphasis on the ecosystem approach.

3. R. T. Oglesby, C. A. Carlson, and J. A. McCann (eds.), River Ecology and Man (New York: Academic Press, 1972). A multidisciplinary discussion of river ecosystems with the emphasis on man's impact.

How we can enjoy the material comforts provided by technology without permanently destroying the natural environment that is vital to our very existence?

Granville H. Sewell believes that a synthesis of technical skills and insightful wisdom by individual citizens, governmental agencies, and industries can maintain a quality environment. In this book, he explains the nature of major environmental problems—land use, water, air, noise, solid waste, and pests— and examines the possible administrative and technological means of coping with environmental degradation. He analyzes the various objectives for environmental programs, explains existing control technologies, describes the laws and govern- mental agencies involved, and evaluates the array of strategies that can be applied. While cautious about the application of elaborate technologies, he explicitly accepts mankind's depen- dence upon tools both to correct the damage previously wrought and to create more satisfying conditions for human society. Thoughtful in tone, practical in approach, and creative in its solutions, this book sets forth valuable guidelines for everyone concerned about the quality of our lives now and in the future.

Prentice-Hall, Inc., Englewood Cliffs, New Jersey

Jacket photographs:
Elliott Erwitt from Magnum, and SCS from Monkmeyer.

0-13-283

4. Kenneth E. F. Watt, Systems Analysis in Ecology (New York: Academic Press, 1966). An early example of an analytical approach that is becoming increasingly popular.

Numerous excellent textbooks and references on microbiological ecosystems exist. A recent example is:

5. Ralph Mitchell, Introduction to Environmental Microbiology (Englewood Cliffs, N.J.: Prentice-Hall, Inc., 1974).

Chapter 3

LAND-USE PLANNING

> "Oligarchy: A government rest-
> ing on a valuation of property,
> in which the rich have power
> and the poor man is deprived
> of it." Plato (c. 400 B.C.)

Role of Land

Manifestations of Problems

Land-use planning is the conceptualizing, coordination,
and encouragement of private and public use of land to satisfy long-term public interests. Land provides the spatial
dimension of our communities and underlies--figuratively and
literally--most of our environmental problems. Dispersal or
concentration of pollutants depends upon spatial location of
activities. Urban and rural aesthetics rest upon the use
or misuse of space. Ecological disruptions occur when we
use spatially-located resources to satisfy economic, social,
and recreational needs.

The problems are being defined in the public mind by a
combination of personal observations, release of public reports criticizing urban sprawl, and the spread of a folklore
questioning a life style that has accompanied some of our
land use. Urban dwellers have witnessed the inexorable
march of suburbs over once-fertile farmlands, woods, and
wetlands. Roads, shopping centers, factories, and office
buildings have leapfrogged apace with the housing developments. Soil has eroded and clogged streams, solid wastes
have accumulated at a quickening pace, rivers have become
polluted from the septic tanks of suburbia, and the utilities and public services have become overtaxed. Even remote
natural areas have suffered from the development of second
homes. Nor has Europe been spared these penalties of affluence, especially in some popular vacation areas, as the
mountain resorts of Switzerland, Italy, and Austria.

Another disquieting development has been the emergence
of reports challenging the long-held tradition that community

growth can be expected to lower taxes by distributing governmental costs over a larger population. Studies, such as The Cost of Sprawl, released by the Council on Environmental Quality in 1974, indicate that the costs of serving a larger population with the prevalent low-density settlement pattern may exceed the taxes that can reasonably be collected.*
Some communities are even finding high-density housing a burden when new schools, water treatment facilities, public transportation, and the full range of urban services are considered. Meanwhile, Pete Seegar continues to sing the broadside, "Little Boxes," satirizing suburban life; and movies, such as The Graduate, raise questions about the apparent amenities of suburban living.

In the last few decades, land-use planning and controls have been increasingly used to protect valuable resources, including existing developments, and to meet the emerging human needs in an economically efficient but psychologically satisfying manner. The problems extend beyond simply passing zoning ordinances or organizing planning commissions, however, and include how to define quality and the role of quantity. To change our controls over land, fundamental changes have to be made in popular attitudes and many of the institutions of local government.

Causes

Difficulties over controlling land use are essentially cultural in nature but are reinforced by social, economic, and political institutions formulated over several centuries. Traditionally, land has been viewed in the United States as a commodity to be bought, used, and sold at will. It has been the legendary source of vast wealth, a reputation not undeserved. As a parlor game and a way of life, Monopoly has been an American favorite. Both governments and private citizens have contributed to this tradition. For example, the federal government tried to finance construction of the Capitol and the White House in Washington, D.C., by sale of lots in the District of Columbia during 1791-1792. There were constant complaints about buildings being placed at intersections of avenues and streets, which were envisioned

----

*Real Estate Research Corporation, The Cost of Sprawl: Detailed Cost Analysis, A Report for the Council on Environmental Quality, The Department of Housing and Urban Development and the Environmental Protection Agency (April, 1974).

as open spaces in the original plan. Philadelphia, New York, and other cities suffered similar problems. In other words, the quality of land and its potential contribution to the greater society's welfare have too often had lower priority than financial gain through speculation and land exploitation.

With limited land resources and a vastly larger population than earlier growth periods, the United States today appears to be in a period of enormous urban expansion. Because of the population boom following World War II, the formation of new households has been increasing as the children of that era begin to bear a new generation. If the current low birthrate continues, the U.S. population can eventually become stable; but this cannot happen for at least four or five decades. Meanwhile, the shift of population to urban regions continues. Suburbs are expected to absorb more new residents between 1970 and 2000 than moved there between 1950 and 1970. The total urban population is expected to more than double between 1960 and 2000. In other words, more than twice the number of residential units, schools, shopping facilities, and utilities will have to exist by the year 2000 than existed in 1960.

Other trends tend to potentially intensify the problem. Despite recessions and inflation, the real per capita income of the U.S. population continues to rise, providing resources for financing development. New technologies, such as land-clearing machinery or building-construction techniques, continue to improve, lowering the effective cost of the development. Affluence and mobility have enabled Americans to intensify use of urban and environmental facilities, especially recreational. Political and economic institutions provide a convenient inertia against change.

Local communities--the municipalities and counties--are most directly affected by land-use decisions; and, as a consequence, controlling land use has traditionally been viewed as a local problem, although the power of land-use control rests constitutionally with the states. (Regulatory powers have generally been delegated by state law to the local communities.) Local governments, though, are vulnerable to parochial pressures. Real estate agents attend all the pertinent meetings and exert powerful personal pressures for favorable decisions. Voters often respond to arguments that appear to promise expansion of the economic base by inviting industrial or commercial investment.

The tax structure, though, has been used as the whip to keep away innovative ideas that might threaten investment in its familiar forms. The property tax, based on appraised value of the land and any structures, is the traditional means of funding education and local services. This means that every local government is under intense pressure to maximize "ratables," high-valued properties that can be taxed accordingly, and lower taxes on other residents. This has placed local governments in a competitive situation with each having to outbid others for the more attractive investments. In this setting, parks, woodlands, wetlands, and farms become viewed with disdain because they represent "unimproved" (low tax-revenue producing) lands.

Countertrends exist. Urban planners have noted with relief a series of recent state court judgments attacking the financing of education solely by local property taxes when this creates inequities between communities. The impact of high-cost energy, especially upon travel and space (interiors of residential and commercial buildings) heating, will undoubtedly be to encourage more compact development and an emphasis on public transportation with clustered development. But the most potent force for change is the uneasy shift in public values against uncontrolled land use and the environmental damage that has been produced in the past.

Mood for Change

Public willingness to support institutional change is difficult to measure and, some critics charge, just as volatile and unreliable. However, events in at least three areas--advisory reports, legislation, and court decisions-- indicate that the public debate that usually accompanies institutional change in the U.S. is occurring. Although perceptive critics of our land-use attitudes have been expressing views for several centuries, the appearance of these opinions in advisory reports, official and unofficial, for high governmental circles is recent. Since 1970, the annual report, Environmental Quality, of the Council on Environmental Quality has been citing the relationship between environmental degradation and land use. In 1970, the Council issued a study, The Quiet Revolution in Land Use Control, which summarized and evaluated land-use laws that had appeared over the country, including Hawaii, Vermont,

San Francisco, Massachusetts, Wisconsin, and Maine.* Outside of official circles, the Rockefeller Brothers Fund
financed a task force and subsequent report entitled The
Use of Land: A Citizens' Policy Guide to Urban Growth,
which strongly supported increased controls.** At the
federal level, numerous efforts were made in the early
seventies to pass national land-use policy legislation
through Congress, but all fell victim to lobbying by various interests or the fear among western states of disruptions that would be caused if the huge federal landholdings
were managed differently.

Indirectly, though, the Federal Clean Air Act Amendments of 1970 are expected to have an indirect land-use
impact because the Environmental Protection Agency is
authorized to (1) oversee implementation of transportation
control plans needed to attain the new ambient air-quality
standards, (2) approve construction of facilities, such as
sports stadiums, that attract traffic, (3) prevent "significant deterioration" of high quality ambient air, (4) define
new source performance standards (the amount of a pollutant
that factories or other stationary pollution sources can
emit), and (5) require maintenance of air quality in metropolitan areas by implementation of ten-year plans. An ap-
prehension frequently being voiced is that those requirements will tend to encourage dispersal of investments away
from centers of population, increasing urban sprawl; but
others note that energy costs and similar trends may be
providing a counter influence.

Because of an emphasis on control of polluting sources
rather than ambient standards, the Federal Water Pollution
Control Act Amendments of 1972 do not pose the same potential problem, although three provisions may encourage closing of some plants and construction of new ones. Industries
must (1) use the best practicable or best available pollution control technologies, (2) pay the full costs of treating wastes discharged to municipal plants, and (3) pretreat
wastes before discharging them into municipal systems.

---

*Fred Bosselman and David Callies, The Quiet Revolution in
Land Use Control, A Report Prepared for The Council on Environmental Quality (December, 1971).

**William K. Reilly (ed.), The Use of Land: A Citizens'
Policy Guide to Urban Growth (New York: Thomas Y. Crowell
Co., 1973).

52

Control technologies are often less expensive if a new plant can be designed and constructed with that purpose in mind than trying to add control devices to old equipment.

States have not been as inhibited as the federal government. The trend to reclaim from local government some of the inherent powers of the state in land-use planning and management has spread until about forty-eight of the states have recently passed legislation expressing the same objective to differing degrees. Much of the legislation is drawing heavily from a draft model land development code prepared by the American Law Institute.

The trend in the judicial system is less consistent, mainly because of the clause in the Fifth Amendment to the U.S. Constitution that states ". . . nor shall private property be taken for public use without just compensation." Known as the "Takings Clause," the concept can be traced to an article in the Magna Carta of 1215 that insured the King would not seize lands or other properties of the subjects. Interpretations of the concept have subsequently diverged.

Under the evolving British tradition, the land and its use are considered an integral part of the nation, so the society retains a vested interest in insuring that land use enhances, not detracts from, the society's welfare. This tradition has two implications. (1) Individuals may hold land ownership and use it for traditional purposes, but regulation to protect the society from undesirable land-use changes is accepted as normal. If the regulation prevents the owner from developing the land's potential for higher monetary value, that is not the society's responsibility, especially when most increases in potential value are due to public investments. If the land were acquired by the government, of course, the owner would be given full compensation. Otherwise, the owner would be eligible for compensation only if regulation prevented any "reasonably beneficial use." Even if an owner believes this has occurred, he would still be reluctant to file notice for compensation because the payment is on the basis of land without "reasonably beneficial use."

(2) Regulation is also more easily accepted because public planning of land use has a stronger and longer tradition. Every significant land development requires planning permission, which is only granted if it complies with the local authorities' development plan. Negotiations and appeals are possible if disagreement exists. Because this

interpretation encourages stability, the British have been able to preserve environmentally desirable features, such as the large greenbelts about major cities, without having to pay immense compensation or engage in protracted litigation.

While the U.S. courts have also recognized this tradition, the dominant theme for the past half century has been that social gain must be weighed against the economic loss of the owner. Compensation should be based upon any "diminution of value," which can include speculative expectations. Yet reasonable expectations are not developed by an extensive planning tradition that clearly establishes social interests beyond immediate political whims. Nevertheless, courts in some states have been supporting stricter land-use regulations without requiring compensation, particularly where the value of exceptional environmental quality has been widely accepted by the public and strong legislation has been passed.

Planning Process[*]

Setting Objectives

Land-use planning follows a procedure similar to environmental problem solving except that the situation is generally more complex and the methodologies have been more standardized and perfected. Because of the political significance of land-use planning, public involvement is essential. This must create widespread political support for conclusions or implementation will fail. Intelligent and skilled political leadership may prove more important than technical expertise in producing a satisfactory outcome.

Several types of land-use plans are possible, depending upon the local problems, objectives, and political structure. Planning can consist of establishing a set of policy statements similar to the National Environmental Policy Act of 1969 that will be enacted by the local legislative body to guide all future decisions concerning land use. (Some type of enforcement mechanism, such as the impact statement

---

[*]The purpose of this chapter is to familiarize readers with the general procedures followed. For more details on preparing land-use plans, see some of the references described in Suggested Readings at the end of the chapter.

requirement, is usually included.) Or detailed maps can be
prepared indicating the location of future activities in
every economic and social category. The most common land-
use plan today is probably a combination of both approaches
with some general statements of policy accepted by the
political system and some concrete, but not too detailed,
physical plans to guide the future decisions. Some of the
detailed objectives that can be considered are the level and
type of economic growth desired, the features of the physical environment that are to be preserved or changed, facilities that are needed to enhance desired social environment,
and the relationships that will exist between the community
and the region.

Base Studies

In urban growth, the future evolves from the present;
and a thorough understanding of the existing land-use pattern and its potentials is essential for preparing a reasonable plan. Because of the complexity of land use, numerous
aspects can be studied in a series of surveys and analyses
known as the base studies. Physical characteristics, economic activities, environmental quality, land value, open
space distribution, hydrological conditions (including flood
plain), cost and revenue flows, socio-economic characteristics, population distribution, and development preferences
are some of the important considerations. For each particular situation, however, a planner has to exercise judgment
on the studies needed to meet his particular objectives
within the financial means available for the overall task.
One would not use a canoe to carry an elephant across a
river but neither would one seek an ocean-going freighter.
Something appropriate to the needs is used.

Almost all land-use studies have, as a minimum, two
base surveys: physical and economic use. Most will find
some socio-economic information extremely useful, if not
necessary. This does not mean that other aspects are not
considered, however, since the planners are generally
assumed to be sufficiently aware of their communities'
characteristics that they know many characteristics, such
as lines of informal communication.

1. Physical Features. Like the skeleton of an animal
limits the function of its limbs, the land's topographical
elements--slopes, streams, lakes, hills and valleys, and
large man-made engineering works--dictate to varying degrees

the human developments that exist or will be added in the future. Three sources are used to study physical aspects and the general pattern of existing development: existing maps, personal survey, and aerial photographs. Nearly every area of the United States is covered by numerous maps compiled for different purposes. Many will have maps prepared by previous planning efforts. Towns and counties have engineering survey maps showing public investments, such as sewers and roads. Court houses have plats (property maps) showing property lines. Highway maps have been prepared by highway departments and, on a cruder scale, are available commercially for community use. The U.S. Geological Survey has topographical maps covering the entire United States, many with a scale of 1:2400 (one inch on map for every 2,400 inches on ground) and lines showing twenty-foot contours (lines of equal elevation). Most studies would eventually have to acquire a base map, one that is clear except for significant landmarks, to record information that will be useful in later stages of the study. Often a grid is superimposed on this base map to provide reference points and facilitate organization of information.

Personal surveys can quickly verify the accuracy of information contained on existing maps. The most versatile tool developed in recent decades, though, is the aerial photograph. Properly used, the photograph will not only indicate the current extent of physical features but will provide qualitative information as well. Types and density of vegetation, texture of surfaces, and even vertical displacements can provide clues unobtainable from standard maps and, sometimes, ground surveys.

2. <u>Economic-Use Survey</u>. Identification of the economic or functional use of land and its potential uses in the future is critical. Because economic-use surveys have been conducted so frequently, standard methods of recording the information have evolved. Data on the type of activity, the land area involved, and--frequently--some qualitative indicators, such as the number of persons or economic value are assembled from existing maps and personal surveys. Categories of use would include residential, retail business, transportation and related activities, industrial and related activities, wholesale business and related activities, public buildings and related activities, open space, and institutional buildings and areas. When entered on a map, each use is given a distinctive color. When tables are compiled, some reference is usually maintained on the grid location of each activity.

Because this information can serve a multitude of purposes and may be required by future studies, a standard method of classification is recommended. In 1965, the U.S. Urban Renewal Administration and the Bureau of Public Roads issued the Standard Land Use Coding Manual, which has been widely adopted. Essentially, this uses a system of digit categories partially based on the Standard Industrial Classification (SIC) used by numerous federal agencies, including the Bureau of Labor Statistics, Bureau of the Census, Social Security Administration, and the Bureau of Employment Security. All activities have been divided into nine major categories. To provide more details within each major category, another digit is added, thus producing 67 two-digit minor categories. By continuing to add digits, 772 four-digit sub-categories are provided. See Table 3-1 for the two-digit classification through the first two major categories.

Sometimes communities prefer to adopt a system being used in neighboring metropolitan areas or by a predecessor planning group. One widely used example is the Land Use Classification Manual devised by the Detroit Metropolitan Area Regional Planning Commission.* If the information is coded by some standard system, it can be placed in a central data repository. Anyone using central repository data, though, does so at some risk. Such data becomes outdated quickly, and there will be frequent questions about quality unless the individuals performing the survey are known and their purpose in collecting the information is clear.

3. Socio-Economic Information. Some general socio-economic information, such as average incomes, can be obtained on a communitywide basis from census data. Since the census is performed only once each decade and local details are often not provided to prevent release of personal information, interview surveys will frequently have to be undertaken. These can provide a multiple service, however, since a competent interviewer can use the opportunity to acquaint households of the planning activity. As views are registered and public consciousness is raised, the possibility of generating political support for eventual recommendations improves.

---

*Detroit Metropolitan Area Regional Planning Commission, Land Classification Advisory Committee, Land Use Classification Manual (Chicago: Public Administration Service, 1962).

Table 3-1. A Standard System for Identifying
and Coding Land Use Activities--
One- and Two-Digit Levels

Code	Category	Code	Category
1	Residential	11	Household units.
		12	Group quarters.
		13	Residential hotels.
		14	Mobile home parks or courts.
		15	Transient lodgings.
		19	Other residential, NEC.
2	Manufacturing	21	Food and kindred products-- manufacturing.
		22	Textile mill products-- manufacturing.
		23	Apparel and other finished products made from fabrics, leather, and similar materials--manufacturing.
		24	Lumber and wood products (except furniture)-- manufacturing.
		25	Furniture and fixtures-- manufacturing.
		26	Paper and allied products-- manufacturing.
		27	Printing, publishing, and allied industries.
		28	Chemicals and allied products--manufacturing.
		29	Petroleum refining and related industries.

Source: U.S. Urban Renewal Administration and Bureau of
Public Roads, Standard Land Use Coding Manual
(Washington, D.C.: U.S. Government Printing
Office, 1965).

Forecasting Needs

The standard procedure for estimating land needs is to select some indicator of growth, usually population, although retail sales or another statistic is sometimes used, and to project future growth by indicator's past behavior and anticipated changes in conditions, such as resource scarcity and new economic activities. If growth over the past twenty years had been consistently 4% annually and no changes in conditions could be anticipated, the future growth rate would be assumed to be about 4% annually, or a total increase (compounded) of 219% in twenty years. Then the projected indicator is multiplied by a standard land-use requirement to produce the land area needed.

Difficulties with this procedure should be obvious. As explained in Chapter 1, forecasting the future is risky, especially when the forecasts involve a phenomenon as volatile as human migratory behavior. Added to this risk is the arbitrary and vague nature of standards. (The standards most widely used in urban planning were developed by a committee of the APHA--the American Public Health Association.[*] Other national standards do exist, such as those assembled and published by the U.S. Bureau of Outdoor Recreation.[**]) When two risky figures are multiplied, the product is an even more risky--and challengeable--number to use as a basis of planning.

Though faulty, this procedure of estimating needs is still superior to basing numbers on simple whims. Rather than using national standards, however, a more defensible procedure is to thoroughly understand local conditions and to derive standards based on these. National standards can then be used simply as a crude check. As an example of an APHA standard, the suggested acreage in parks (local) for a neighborhood of 3,000 persons containing one- or two-family dwellings would be 2.5 acres, assuming private lot areas of less than one quarter acre per family. By comparison, the National Recreation and Park Association recommends a minimum of about 15 acres (active recreation) for 3,000 people

---

[*] Committee on Hygiene of Housing, American Public Health Association, Planning the Neighborhood (Chicago: Public Administration Service, 1960).

[**] U.S. Bureau of Outdoor Recreation, Outdoor Recreation Space Standards (April, 1967).

located within an urban area plus another 15 acres in
natural or landscaped parkland.

Preparing the Plan

Plan preparation consists of devising a strategy for
satisfying future land needs from anticipated resources.
As noted earlier, the plan can consist of a simple policy
statement, a detailed land regulation program, or both. In
the process of any detailed analysis, the qualitative nature
of the resources must be examined. For example, industries
prefer nearly level land; access to highways, railroads, and
airports; safe water supply; satisfactory methods of waste
water disposal; employment pool; and some amenities for the
employees.

Allocations of available land have to be gradually made
to projected needs. This ideally would be the time for an
intensive public information campaign to publicize the
issues and generate political support for decisions as they
evolve. In practice, this stage becomes difficult to handle
because land speculators are quick to take advantage of any
advance information on land regulation that would enhance a
property's value.

Land-Use Controls

Choices

The purpose of land-use controls is to influence the
location of future investments in order to comply with the
land-use plans. To exert this influence, a planner has four
categories of tools. (1) The investment can be discouraged
directly by prohibiting certain developments or indirectly
by limiting activities needed for the developments to function. This is the stick approach, admonishing developers
that "thou shalt not deviate from what I say." (2) Desired
investments can be encouraged through selective application
of governmental powers. This is the carrot approach, the
tax preferences or construction of highways that lower development costs. (3) Development can be preempted by the
government acquiring development rights. Thus, the planner
representing the society becomes the developer and can blame
no one if plans go awry. (4) A hybrid or high threshold
approach requiring potential investors to participate in the

planning process can be used. Only developers capable of
mustering the planning and endurance capabilities to overcome this obstacle can make the investments.

The Stick

Zoning, the prohibition of certain developments or
activities in designated areas, is undoubtedly the most
common deliberate control wielded over land use. In order
not to discriminate against certain neighborhoods, the entire town or county must be divided into development zones,
such as single-family residential, light industrial, and
commercial. Each zone is described in an ordinance that
states the zoning objectives and lists criteria for issuing
development permits. In its simple form, zoning is protective and perpetuates the status quo by turning away any
developments outside the ones described by criteria.

As a technique, zoning has several glaring faults.
Because land values will differ between zones, an incentive
will exist for developers to evade the restrictions. A
store in the midst of a residential area, for example, would
have a form of monopoly. An even more glaring example would
be a private dwelling set in the midst of a park, a situation that would raise the value of the dwelling but lower
the value of the park--and adjacent property--for the rest
of the community. Developers, therefore, are constantly
trying to obtain variances, permission to waive zoning criteria. The process is known as spot zoning and is usually
prohibited by law.

A more fundamental difficulty is zoning's inherent
flexibility. It assumes separation of functions. Why not
locate a few light industries and commercial establishments
in a residential district? Transportation problems for
employees may be reduced, and the design aesthetics could
be pleasingly varied. One successful device for overcoming
this difficulty in some cases has been the creation of
special districts with the zoning criteria being preservation of unique social, cultural, or historical values. All
development plans can then be reviewed with this objective
as the criteria.

Investments can also be discouraged by indirect devices. Prohibiting connection of new structures to sewers
is tantamount to prohibiting building since no one wants
to be flooded in their own sewage. Limiting the size of

parking lots effectively discourages large shopping centers and some industries. Deliberate timing of administrative procedures can also limit or discourage growth. For instance, a governmental office that issues building permits can establish a policy of issuing only a certain number of residential unit permits per month.

The Carrot

Government initiative can effectively stimulate certain types of development. For example, urban sprawl in some areas of the country has been encouraged by the government financing of extensive road and sewer construction. Citizens or developers who exhibit desirable behavior can be rewarded by tax reductions in the form of preferential assessments. Over thirty states have already adopted some form of preferential assessment for property taxes to protect forests, farm lands, recreational resources, and other special purposes. (To prevent speculators from using this device as a tax shelter, some high penalty is usually attached to any efforts to change the land use.)

Preemption

European nations have enjoyed a high degree of success in moves to preempt development by acquiring development rights over the land, and the costs have usually been regained by later sale of part of the lands after they have appreciated through government investments. In the U.S., however, the general poverty condition of local governments plus the lobbying efforts of speculators have usually blocked this option. A few innovative efforts have been made, though, by some agencies, such as the U.S. Corps of Engineers and the quasi-public Urban Development Corporation of New York State.

Two approaches are used. The land and all of its rights can be purchased outright. If the land can be used by the government, as in the case of parks, the ownership will be retained. If use or future use is not contemplated, the use rights can be leased or sold to a private party or other governmental unit. For example, farmland can be purchased and the grazing or farming rights leased or sold. Future development of this land without the cooperation of the government would then be impossible. In the second approach, only the necessary development rights are

purchased, as in the acquisition of easements along a river
to prevent erection of advertising boards or to permit access by the public. Sometimes these development rights are
donated to the government to protect land values for residential communities or the rights will be pooled into a
cooperative or corporation with the express purpose of protecting the community from certain types of developments.

High Threshold Strategy

Instead of assuming that the public sector must determine the acceptability of a developer's program, the planners can take a different stance and insist that the developers take the initiative in proving acceptability by
detailed analyses of the development's impact. In other
words, a developer's plans have to be accompanied by analyses akin to the environmental impact statement. This approach is usually called the PUD for Planned Unit Developments. The community still bears considerable responsibility
because a general development plan should exist for guidance.
Talent and available time by local planners should be available to guide and interact with the developer's representatives. Costs of development are raised, and some firms will
not be able to compete. However, this is increasingly viewed
as a preferred approach for protecting the community's welfare.

Implications for Future

Rapid urban development is now occurring in a society
acutely aware of resource scarcity. Energy costs are high.
Material costs are rising. Quality environmental resources
are threatened. In each case, the use of land is a critical
factor. United States society is traditionally averse to
planning, particularly in as politically sensitive an area
as one that affects property values and community identities.
Yet to cope with the inherent dangers of inevitable change,
planning is proving a necessity. The pattern of the planning process will necessarily be unique to the political
conditions, which means that land-use planning in approaching decades will be in a creative atmosphere containing both
the challenges and the rewards of fundamentally shaping
environmental quality.

QUESTIONS FOR DISCUSSION

1. Why are open space sites, such as parks, often proposed for location of public facilities, including schools and roads?

2. Explain the ways that the automobile has influenced our use of land.

3. From (a) ethical and (b) functional points of view, should steps be taken to discourage or encourage land speculation?

4. If you were to use land-use planning for improving air quality conditions in your community, what specific recommendations would you make? For water? Noise?

5. Discuss the differing functions you would expect between a local and a regional land-use planning effort.

6. Other than total population, what characteristics, such as sex, of a forecasted population would be useful in understanding a community's future needs?

7. If you had made a population forecast of your community twenty years ago, what events have subsequently occurred that could not have been foreseen and yet would have changed your assumptions?

SUGGESTED READINGS

In the context of this chapter, important readings can be placed into three categories: planning textbooks, reports or commentary on land use, and classics on urban themes. Everyone in environmental management should be familiar with the contents of Principles and Practice of Urban Planning.

1. William I. Goodman and Eric C. Freund (eds.), Principles and Practice of Urban Planning, published for the Institute for Training in Municipal Administration (Washington, D.C.: International City Managers' Association, 1968). The urban planning field as it is envisioned by leading practitioners is discussed concisely, authoritatively, and clearly. In one sense, this book has shaped recent evolution of the field since it is the recommended

manual for preparing to take licensing and other examinations in urban planning.

In the more specialized field of land-use planning, several leading books are:

2. F. Stuart Chapin, Jr., Urban Land Use Planning (Urbana: University of Illinois Press, 1968). The standard reference for urban and regional planners with a physical, land-planning orientation.

3. Charles Haar, Land Use Planning (Boston: Little, Brown and Co., 1959). The standard authority on land-use planning from a legal and policy point of view.

Between being a textbook and a report is The Use of Land, edited by William K. Reilly, a lawyer-planner. While representing a committee's report, the writing has a clarity and usefulness that rises above that normally expected from a committee. The other reports listed below were discussed earlier in the chapter.

4. William K. Reilly (ed.), The Use of Land: A Citizens' Policy Guide to Urban Growth, a task force report sponsored by the Rockefeller Brothers Fund (New York: Thomas Y. Crowell Co., 1973).

5. Real Estate Research Corporation, The Cost of Sprawl: Detailed Cost Analysis, a report for the Council on Environmental Quality, The Department of Housing and Urban Development and the Environmental Protection Agency (April, 1974).

6. Fred Bosselman and D. Callies, The Quiet Revolution in Land Use Control, a report for the Council on Environmental Quality (December, 1971).

For more leisurely reading, a vast library on urban planning topics exists.

7. Sigfried Giedion, Space, Time and Architecture, 5th edition (Cambridge: Harvard, 1967). A classic history of architecture and urban planning. Many pages but provides rapid and fascinating reading.

8. Kevin Lynch, The Image of the City (Cambridge: M.I.T. Press, 1960). A standard authority on importance of urban form. Contains analyses of several urban examples.

9. Jane Jacobs, <u>The Death and Life of Great American Cities</u> (New York: Random House, 1961). Provocative analysis of urban life and design, especially the need and elusiveness of vitality.

10. O. Schoenfeld and H. Maclean (eds.), <u>City Life</u> (New York: Grossman, 1969). A thoroughly entertaining anthology of literate writings on city life.

Chapter 4

SIGNIFICANCE OF WATER QUALITY

"Everywhere water is a thing of
beauty, gleaming in the dewdrop;
singing in the summer rain;
shining in the ice-dipped gems
till the leaves all seem to
turn to living jewels; spreading a golden veil over the
setting sun; or a white gauze
around the midnight moon."
John B. Gough, A Glass of Water
(1817-1886)

"To many New Yorkers, the Hudson River is little more than
an evil-smelling sump whose
water is foul with sewage, oil,
orange rinds, detergents, thick
wads of pulp, and sundry other
human and industrial wastes.
It seems a small miracle that
even the sinister eel manages
to survive on the river's roily
bottom." Business Week (January 31, 1970)

What is Water?

The Chemical Dimension

Pure water is H_2O, a liquid of hydrogen-oxygen molecules. But fortunately water is never pure in nature.
Gases, including oxygen, carbon dioxide, and nitrogen, are
dissolved between the water molecules. Salts, such as
nitrates, chlorides, and carbonates, also become part of the
liquid solution. Solids--tiny bits of animal matter, dust,
and sand--can be carried as suspended solids. Other chemicals give water color and taste. Ions may cause a chemically alkaline or acid reaction. Temperatures will vary with
depth and location, influencing chemical behavior.

The Biological Dimension

Within this complicated mixture exists an extraordinarily varied array of plant and animal life. From single-celled phytoplankton (plants) and zooplankton (animals) to the Blue Whale, the entire spectrum of size is found in water. And almost every class of life outside water finds its counterpart within water. Some, particularly insects, may live different stages of life both within and outside of water. But each category of biota will have a particular niche, or set of water conditions, that it finds most suitable for development.

The Time Dimension

To this description, a time consideration must be added because water in the natural environment is changing constantly according to complicated patterns. Cycles are imposed upon cycles. Water is continually being drawn into the atmosphere as vapor and falling again as precipitation. As much as three-quarters of this precipitation evaporates from the soil, lakes, and streams or transpires from growing plants without being directly usable by mankind. Rainfall varies geographically, and it has seasonal fluctuations, more falling during the summer than the winter, although winter streamflow may attain higher peaks because of floods from melting snow.

As each day--or season--passes, surface temperatures for water change. In a chain reaction, gas levels in the water will also vary since cold water can hold more dissolved gas than warm. Measurements of carbon dioxide over oceans have shown that the atmospheric carbon dioxide falls when colder weather causes the absorption of gases, then rises again with warmer weather.

Gas cycles are complicated by the photosynthesis and respiration by plants since, like animals, they respire by absorbing oxygen and ejecting carbon dioxide but, when exposed to light, they will absorb the carbon dioxide and release oxygen. Thus, dissolved oxygen levels in a stream can fall at night.

The Cultural Dimension

When managing water resources, discussion of water in unemotional, scientific terms alone misses a key element in the human response to the topic. Water has an important value, both materially and emotionally. To farmers, the presence of water at critical times can mean the difference between success and failure. Since fish need water of appropriate conditions, so do fishermen. Cities require water to drink, bathe, and extinguish fires. Industries use water to cool equipment, mix with products, cleanse materials, or simply convey substances. Ships obviously need water, and the shipping industry can be depended upon to lobby on behalf of ship channels. For nearly every economic interest, water represents an inexpensive and convenient "sink" for wastes that cannot often be traced back to their origin. Water is also avidly sought for recreation--swimming, boating, and fishing--and land overlooking water is prized for almost every social purpose.

Part of the emotional response lies beyond simple rational explanations. Fountains are built for the beauty of dashing water. Water has inspired poets, artists, and composers. Children sing about a "rolling Shenandoah" and are told stories about a "mighty Mississippi." In the imagery of the water environment are the paintings of the Hudson River School and individuals like Winslow Homer or Joseph Turner. Biblical quotations of "the still waters" are embedded in our collective subconscious. A quiet pool, a reflecting lake, a majestic river, a storm-swept sea--each phrase suggests a personal relationship with the physical world. This emotional response must be considered in planning for a more satisfying physical environment.

Need for Water Management

Water Uses

Uses of water can be divided into four broad categories:

(1) Direct physical use by man and his domesticated animals. We use water for drinking, bathing, laundry, and various household uses. Standards are stringent. Water must appear clean, taste pure, and--above all--be free of the dangerous and debilitating microorganisms that scourged

urban populations of previous centuries. Surprisingly, less than 5% of the approximately 360,000,000,000 gallons of water drawn daily in the United States serve this use.

Table 4-1. Estimated Water Use: 1940-1970
(Billion gallons average daily use)

Year	Total Water Use	Irrigation	Public Water Supplies	Rural Domestic	Steam Electric Utilities	Industry and Miscellaneous
1940	136	71	10	3	23	29
1950	203	100	14	5	46	38
1960	323	135	22	6	99	61
1970	327	119	27	4	121	56
1975 (est.)	385	128	30	5	157	65
Percent use in 1975	100%	33%	8%	1%	40%	17%

Source: Bureau of the Census, U.S. Department of Commerce.

(2) Direct use in industry and agriculture as a factor in production. Industries are the largest consumers, nearly 40% being drawn for cooling by the electric utilities alone. Another 17% is classified "industrial and miscellaneous" and includes manufacturing, mining, and construction. Irrigation accounts for 33%. Regionally, irrigation is attributed with use of more than 80% of the supply for seventeen western states while industries are the predominant users in thirty-one eastern states using over 80% of the supply.

(3) Psychological use as part of our esthetic and cultural environment. A scenic lake or river may not be serving other purposes (though most are) but, if rational economic accounting occurs, the water has a definite value because satisfaction is being provided the viewer as surely as the taste in his food or the color in his clothes. The value cannot be measured neatly in dollars.

(4) Ecological use as a vital component in the earth's life support system. Despite the earth's vast volume of

water, which would flood the land to a depth of about 800 feet if distributed evenly, the water environment is biologically fragile. Only about 3% is fresh water. In the remainder, biological production is limited primarily to the surface layer, especially in estuaries. Yet the entire ocean volume acts as a heat and mineral thermostat for the rest of the world.

Water Distribution Factor

For all uses, water rarely exists in the quantity, quality, and location that humans would consider ideal. Total stream flow in the United States is about one trillion gallons per day. Much of this sweeps unused to the sea during occasional floods. Yet we were using about 327 million gallons per day on the average during 1970, and use will exceed 400 billion gallons by 1980 according to U.S. Department of Commerce estimates. Rivers in many areas, such as the South and Northwest, are not heavily used. This means that some rivers, such as the Hudson, Ohio, and Delaware, in the industrial, heavily populated areas of the country, are being used four or five times their average flow--without counting their use as a sink for wastes.

Present Water Quality

By Environmental Protection Agency's estimates in 1970, almost one-third of the U.S. stream miles are characteristically polluted in terms of violating federal water quality standards. Less than 10% of the U.S. watersheds are considered unpolluted or even moderately polluted. In the Northeast, about 44% of the stream miles are classified as polluted.

Because data are lacking, we do not know definitely whether average river and lake quality has deteriorated over recent years, but the general assumption is that deterioration has been rapid, perhaps doubling by common indices in the decade between 1960 and 1970. Industrial wastes are now considered more destructive than sewage from municipalities, although the distinction is not completely valid since many municipal effluents also contain industrial wastes. But the "stick and carrot" approach of concerted governmental pressures against municipalities coupled with subsidies for treatment facilities have begun to diminish municipal pollution of some types.

Table 4-2. Estimated Extent of Water Pollution: 1970

Region	Percent of stream miles polluted	Percent of watersheds polluted			
		Over 49%	20-49%	10-19%	Below 10%
Total U.S.	32.6	23.7	48.5	17.7	9.9
N.E.	43.6	36.1	55.6	5.6	2.8
Central	36.6	23.2	51.8	21.4	3.6
S.E.	23.3	14.3	41.1	16.1	28.6
S. Plains	38.8	27.3	51.5	18.2	6.1
N. Plains	40.0	37.5	33.3	25.0	4.2
Pacific Coast	33.9	14.8	59.3	22.2	3.7

Source: Environmental Protection Agency, Water Quality
        Office.

Agricultural wastes have been becoming increasingly
serious because of fundamental changes in agricultural
practices, although recent changes in energy and fertilizer
prices may be altering the trend. Instead of allowing cattle to roam freely over pasturage where decomposers in the
soil could digest manure, cattle were being raised in restricted space of huge feed lots. Since a cow discharges
about ten to twelve times the organic wastes of a man, a
feed lot of 50,000 head of cattle will be equivalent to a
city of 600,000 population with most of the wastes flowing
into streams during rain. In addition, intensive artificial
fertilization of fields is sending "slugs" of nutrients into
streams with rain; and the trend towards convenience foods
by housewives means more processing--and waste--that will be
deposited in rural processing centers.

Measurements of Water Pollution

"Water pollution," like "evil," can cover a multitude
of sins. In essence, it commonly means that one or more
uses of the water has in some way been impaired. Usually,
"pollution" in water refers to one or more of six common
conditions.

(1) <u>Loss of Dissolved Oxygen</u>

<u>Significance</u>. The dissolved oxygen in the water has been unnaturally diminished or exhausted. The oxygen level is critical. All higher forms of animal and plant life respire, inhaling oxygen and exhaling carbon dioxide. Without oxygen, they die. Oxygen also allows the <u>aerobic</u> (oxygen-using) bacteria, which are more efficient decomposers than <u>anaerobic</u> (nonoxygen-using) bacteria, to reduce decaying organic matter in the water without obnoxious odors.

Oxygen is most often removed by the very efficiency of nature's cleansing mechanisms. When large quantities of organic materials are discharged into streams, a population explosion occurs in the bacteria and other decomposers. They "breathe" oxygen, and the subsequent fall in oxygen level is known as the oxygen sag. If the oxygen is exhausted, the water becomes anaerobic, or septic. An entire bacteria population then has to change. Also, anaerobic bacteria generate hydrogen sulfide, an extremely obnoxious smelling gas that, in sufficient quantities, is toxic.

Fish life is typically killed, but the killing is selective. More sensitive fish, such as trout, are the first to succumb; and the "trash fish," such as carp, can linger longer. Larger and more mobile fish may be able to evade the anaerobic condition if it comes down a river as a slug. But insect larvae typically cannot. Some plants can live for short periods without oxygen, but others cannot. Some shellfish will die immediately, but others may survive for short periods. Oysters, for example, can frequently live as long as twenty-four hours in a quiescent state with shells closed while waters are anaerobic. After such an episode, however, the natural equilibrium between organisms will be disturbed, and years may pass before the full variety of biota returns and equilibrium is restored.

<u>Measurement</u>. Oxygen in water is usually measured by an electronic probe that balances a current with the passage of oxygen through a plastic membrane. The oxygen levels can also be determined chemically by mixing a water sample with certain compounds, such as a manganese salt, and measuring the precipitate.

Equally important is the prediction of the oxygen depletion that some known organic wastes will cause in a stream. This is done by measuring the oxygen-absorbing

73

capability of bacteria and other microorganisms that will decompose the organic matter in the wastes. By some simple arithmetic comparing the oxygen-absorbing potential of wastes entering a stream with the dissolved oxygen available in the stream, a crude forecast can be made of the probable effects. This oxygen-absorbing potential is expressed by several measurements, the most common being Biochemical Oxygen Demand, or, as it is usually called, BOD. The higher the organic content of the water, the higher BOD levels will be. The standardized BOD test is the measurement of oxygen remaining in a bottled sample of water (often diluted) that has been stored in the dark at 20° C. for five days.

The resulting BOD estimate for the wastes in the water is typically expressed "pounds of BOD," the weight of the oxygen that can be expected to be absorbed by the total amount of wastes. To avoid the five-day wait for the BOD test, analysts sometimes use a Chemical Oxygen Demand (COD) test, which simply measures the amount of organic (carbon) materials in the sample. This, though, is nearly always higher in value than the BOD test since all organics are not typically oxidized at once in nature.

(2) Pathogenic Contamination

The water may be infected with pathogenic (disease-causing) organisms. From innumerable historical outbreaks of typhoid fever, cholera, various dysenteries, and other illnesses unmistakably traced to infected water, we know the threat is real; and water-borne diseases still represent major killers in the developing world. The causes of these diseases can be attributed to five categories of parasitic organisms: bacteria, protozoa, worms, viruses, and fungi.

(a) Bacteria. Bacteria are the most common with typhoid fever and cholera among the best-known representatives. Cholera is a violent illness of one to three days duration with a fatality rate in epidemics of 5 to 75%. Typhoid fever takes several weeks to reach a climax and typically has a fatality rate of about 10%. However, typhoid fever is more persistent and produces human carriers who can unwittingly continue to spread their illness. Classic examples exist where carriers have been employed as food handlers; and a spectacular case in England during the thirties was caused by an infected workman who contaminated a well, thus causing 341 cases of typhoid fever with 43 deaths. Other examples of bacteria-caused diseases would be paratyphoid (salmonellosis), bacillary dysenteries (shigellosis), tularemia, and Weil's disease.

(b) Protozoa. Protozoa-caused infections are usually limited to amoebic dysentery (amebiasis), but even this is rare in the United States despite the estimate that between 1 and 10% of the population carry amoebic cysts.

(c) Worms. Intestinal worm, eggs, and larvae are relatively large organisms, and the problem is rarely attributed to water in the United States. But in many developing countries, virtually the entire population is host to some form of worms. Even in Hong Kong during the sixties, drinking water from a hotel tap was an almost guaranteed way to acquire intestinal worms. Importance of the intestinal worm, though, is overshadowed by a distant relative, the blood fluke (schistosome), that is even more debilitating to a large proportion of the world's population. The blood fluke develops in fresh-water snails, which are becoming more common as irrigation farming is practiced in Africa and Asia. A less dangerous form of schistosome occurs in the United States as "swimmer's itch" (cercarial dermatitis) carried between water bodies by infected water fowl and harbored by snails.

(d) Viruses. At least six virus groups with over 100 different strains are known to be released in water by feces. Yellow jaundice (infectious hepatitis) is also believed to be water-borne but has not yet been isolated in tests. Effects from viruses range from the well-known paralysis of poliomyelitis to gastrointestinal upsets, rashes, respiratory difficulties, and inflammation. But virus-caused epidemics from water sources have not been known to reach the dimensions of the bacteria-caused illnesses.

(e) Fungi. Fungi-caused problems are typically limited to skin rashes and, occasionally, eye disorders.

Measurement. Unfortunately, locating and identifying a water-borne pathogen in a laboratory sample would be like the proverbial search for a needle in a haystack. When necessary, the typical search procedure would be to give water samples to laboratory animals and wait patiently for disease symptoms to appear. Obviously, a better indicator for the presence of pathogens is needed, so we have been accepting the presence of a harmless family of bacteria

75

known as the coliform group as evidence that human wastes--
and presumably, pathogens--may be present.

The average person typically discharges daily about a
trillion coliform bacteria in feces. This is extraordinarily convenient since the coliform are normally hardier than
pathogens, do not typically multiply outside a body, and
they are considered safe to handle. In technical language,
the coliform group are defined as aerobic and facultative
anaerobic, nonsporeforming, Gram (stain) negative, rod-
shaped bacteria that ferment lactose (milk sugar) with the
production of gas within forty-eight hours when maintained
at a temperature of 35° C. (95° F.). If we need an estimate
of the coliform density in a sample of water, we can place a
measured amount of sample water onto a glass plate with a
special sterilized sugar film and heat it to about body temperature. After approximately twenty-four hours, the individual coliform will have multiplied until visible colonies
have been formed and can be counted.

The problem is complicated since man's intestines do
not have a monopoly on coliform bacteria. They exist in
many places, including the soil. Their presence may have
no possible connection with the existence or nonexistence
of pathogens. Thus, we may be condemning some water that is
completely innocent of evil potential. To be slightly more
specific, we can undertake a more elaborate testing procedure for Escherichia coli, better known as E. coli or fecal
coli, that is usually found only in intestines. But even
E. coli has been found in some soils and it can be discharged by animals.

Even when we have completed the procedure of proving
the presence or nonpresence of coliform, the interpretation
of the fact remains. Certainly we want to minimize risk
with drinking water and exclude supplies with possible contamination. The Public Health Drinking Water Standards
(1962) do not allow more than one organism of the coliform
group in 100 milliliters of water, but what happens to swimming beaches where some coliform will be found naturally?
Each city and state establishes its own standards, and these
vary immensely across the country. Increasingly, evidence
suggests that coliform counts do not correlate with disease.
Yet swimmers, especially children, do accidentally ingest
swimming water; and, with the coliform group being the only
feasible indicator of the possible presence of pathogens,
public health officials are understandably reluctant to
waive coliform standards completely.

(3) Presence of Nutrient Salts

Nutrient salts, such as nitrates and phosphates, are released, creating a condition known as eutrophication. Eutrophication today is a term used for the accelerated aging of a lake by wastes released by man. In a more prosaic description, eutrophication of a lake--or river--is occurring when the water becomes choked with algae and weeds, the ecosystem becomes unstable, and periodic problems, such as objectionable odors, tastes, and widespread fish kills, become apparent.

All ecosystems age, and an increased population of producers (plants) is a natural trend in lakes. When lakes first emerge as products of geological action, they are typically nutrient-deficient. Plankton, the microscopic plants and animals that float in open water, are scarce. The clear waters absorb the shorter wave lengths of light that usually are reflected so the water has the deep blue color associated with mountain lakes. As nutrients wash into the water and nutrient-fixation occurs, the phytoplanktons (microscopic plants) begin to increase in number and change in species. Desmids are replaced by diatoms and then by flagellates, and similar green algae. Finally, the blue-green algae--one of the few plants that can fix nitrogen in significant quantities--become dominant. An algae "bloom" occurs with all of the associated nuisances of odors and color. The mass of decaying organic material on the bottom begins to mount, and the lake gradually diminishes in volume until it disappears.

This process typically occurs over geological ages; but, by artificially adding nutrients, mankind has managed to compress the time period. The entire ecosystem rapidly changes. Trout, chub, and whitefish are progressively replaced by bass, perch, and pike; and these are finally succeeded by the less favored carp and sunfish. Meanwhile, a sudden freeze can place a cover of ice on the lake or river while, at the same time, killing a summer-induced growth of algae. As they decay, the water's oxygen is exhausted, and a massive fish kill can result. Or another factor can trigger the algae's destruction, again creating anaerobic conditions. The oxygen sag described in the discussion on dissolved oxygen often has an "echo" sag four or five days later as algae produced in the oxygen recovery stage begin to die under the more normal conditions.

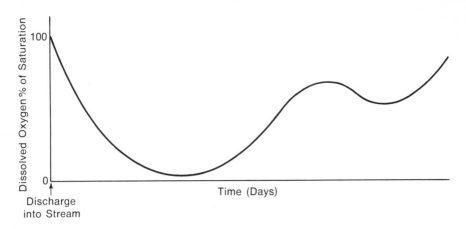

Figure 4-1.   Typical "oxygen sag" for organic waste dis-
              charge into river, assuming full saturation
              of stream before waste enters.

(4) Obnoxious Chemicals or Heat

Chemicals or heat that are toxic or obnoxious to aquatic life--and sometimes man--can be released into water. Except for heat and some of the more unstable organic compounds, these materials degrade slowly--if at all--and must
be either dispersed in immense bodies of water or buried in
the stream or lake sediments. Sometimes they will continue
to pose a hazard for years if the sediment is ever disturbed; and, in the case of mercury, there is a suspicion
that some sediments are continuing to leak toxic compounds
without disturbance.

The list of possible chemical contaminants in water includes virtually every soluble toxic compound that is manufactured or collected by man, including those of the heavy
metals (barium, cadmium, chromium, cobalt, lead, mercury,
and nickel) that are naturally found in trace amounts.
Petroleum spills are becoming more frequent as the nation's
energy demands require transportation of more petroleum.
Salts, especially chlorides, are leached from land by irrigation, released in domestic sewage, and flushed from industrial plants. Pesticides and herbicides are sprayed over
both agricultural land and the urban areas where suburban-
ites are anxious to be rid of nuisance bugs.

In some cases, these materials kill aquatic life immediately; and unless removed by treatment, they could be dangerous in human water supplies. Others, such as organic mercury or pesticide compounds, can be stored in organs or fat tissues of animals until a toxic dosage is accumulated. Sometimes the hazard is more mundane. Excessive salts cannot be used for many industrial purposes because they would corrode the tubes of boilers or heat exchangers or would contaminate chemical reactions. Excessive salts preclude use of water in irrigation. And salts are also the culprit in creating "hard" water that is the bane of many a housewife's existence. Under some circumstances, such as the presence of certain sulfur compounds, the water will be made intolerably acidic or alkaline. In the coal-mining region of Appalachia, some streams have become lifeless because the water is so strongly acid. The counterpart, alkali streams or lakes, can be found in some western states.

(5) Turbidity

Turbidity, the cloudiness of water due to suspended solids, is an example of a condition that can be man-made pollution or a natural phenomenon, depending upon circumstances and the solid. If organic, the solids are often--though not always--from domestic sewerage, industrial processes, or pollution-induced eutrophication. (Of course, some phytoplankton and rotting vegetable or animal matter are necessary and desirable in any productive water.) If inorganic, the suspended solids are usually from the soil, including sands and clays. Again, the question of pollution can only be answered by the degree and place. No one would question the muddy quality of the Mississippi in flood stage, but thousands of miles of high-quality streams have been choked and smothered with sediments, all valuable aquatic life being destroyed, because of soil washed from construction projects, mining, dredging, and insensitive agriculture. Turbidity used to be measured by submerging a colored disc at the end of a calibrated pole, then noting the depth where the disc became indistinct. More recently, electronic devices that use a light source separated by water from a photosensitive cell at a fixed distance have been widely adopted.

(6) Esthetic Insults

Esthetically, water is considered polluted when made unattractive because of visible trash, obnoxious odors, or an unpleasing taste. Coincidentally, these conditions usually indicate other forms of pollution. Visible trash is typically accompanied by invisible sewage. Odors are normally caused by hydrogen sulfide that is generated by anaerobic bacteria or algae from eutrophication.

Precedent for Improvement

If the various forms and the magnitude of water pollution in the United States now appear overwhelming, the achievement of the 19th Century in transforming the urban environment and, at the same time, public attitudes toward water may serve as assurance that near-miracles do occur. Amidst considerable painful debate over reservoir sites, water main routes, and financial details, the major cities gradually developed their first public water supplies during the first half of the century. Then the emphasis shifted to sewers. The first was constructed for Brooklyn, New York, in 1857. Within about twenty years, any city that considered itself significant seemed to have its streets torn apart with gaping holes for sewer installation.

Yet cities were far from pristine. Consider the typical rush-hour scene on a midtown thoroughfare in Manhattan a century ago. Traffic is horse muzzle to carriage back, and swarms of pedestrians trample through ankle-deep mud on the unpaved street. Most houses are still not connected to the public sewerage system, and foul cess-pool overflow dribbles into the street. Swill carts and nightsoil wagons slosh and leak their loads. But it is unnoticed in the mire of horse manure and urine. Odors are nauseating. Rotting, reeking garbage is piled in backyards. Slaughter houses, renderers, and tanneries ply their smelly and messy trades in establishments along the city streets. And dead horses and dogs are still being dumped to putrefy on the river bank a few blocks away. The good old days?

Despite the vagueness of associations between sanitation and health, the public health improvements of the period were spectacular. In early 19th Century America, typhoid death rates ranged from 50 to 175 per 100,000 population annually, and cholera epidemics were even more disastrous, the rate rising over 300 per 100,000. Sewage also provided ample breeding waters for mosquitoes. Malaria was a chronic malady, and yellow fever epidemics constantly ravaged major cities. In the summer of 1793, for instance, someone estimated that over 4,000 persons had died from yellow fever in Philadelphia out of a population of 29,000, a rate approaching 14,000 per 100,000 population. If a comparable death rate from disease were to strike the modern New York metropolitan area, over two million persons would suddenly die. In Philadelphia, ". . . everyone who could afford it abandoned his business and fled through the stricken city. Day and night the death carts rumbled through the town. The streets were as those of a dead city. Life-long friends evaded one another like guilty creatures. Even the families of the stricken fled, leaving the suffering to die in barbarous neglect."[*]

Yet the causes of these diseases remained vague. Among accepted techniques for ending the plagues were ". . . burning nitre in the streets, firing horse pistols at the bedsides of the sufferers and carrying garlic in their shoes and bags of camphor around their necks, and dousing themselves with Haarlem Oil and essence of aloes, and Vinegar of the Four Thieves."[*] Streets were barricaded and vigilante committees patrolled to keep out citizens from stricken communities.

An interesting aspect of the 19th Century experience is the change in public attitude towards water quality. In 1800, pure water was viewed as a desirable commodity, but still dispensable.

> Pure water . . . is the best drink for persons of all ages and temperaments. By its fluidity and mildness, it promotes a free and equable circulation of the blood and humours through all the vessels of the body upon which the due performance of every animal function depends, and hence water-drinkers are not only among the most active and nimble, but also the most cheerful and springly of all people. . . . But to delicate and cold constitutions, and to persons unaccustomed to

---

[*]Quoted in Gilbert F. White (ed.), Water, Health and Society: Selected Papers of Abel Wolman (Bloomington, Ind.: Indiana University Press, 1969).

it, water without wine is a very unproper
drink.*

By 1900, a public supply of pure water was considered
a necessity and right. Similarly, sewage running in the
streets was considered normal for 1800, and even 1850. By
1900 these conditions had become intolerable. Bathrooms
began being installed in houses about 1830. Previously, a
toilet in a house was considered a vulgar, unclean idea; and
a fixed bathtub was a senseless use of space. Yet changes
in attitude in one area begot changes in others. Unexpected
but happy results fueled the demand for even more drastic
improvements.

Possibly the same transition in values, the way of
viewing life's priorities, is evolving in our attitudes
towards man's dirtying the natural environment. Our grand-
children may be able to view our present destructive disregard for natural water quality with the same wonderment
that we regard our ancestors' seemingly primitive and
narrow-minded tolerance.

QUESTIONS FOR DISCUSSION

1. Outline arguments from three points of view--biological,
    economic, and esthetic--against water pollution. How do
    moral questions enter each?

2. Would you expect the effects of two forms of water pol-
    lution--say, BOD and heat--together to be more or less
    than each taken separately.

3. Define the effects that untreated domestic sewage dis-
    charged into a typical, unpolluted river ecosystem would
    have upon aquatic (a) producers, (b) consumers, and
    (c) decomposers.

4. What are the advantages and disadvantages of recycling
    water for domestic supplies, assuming that satisfactory
    forms of treatment exist for each pollutant?

5. Describe an ideal water quality monitoring system for a
    river. What compromises would you expect to make for
    practicalities?

*"On the Means of Preserving Health," Philadelphia Monthly
Magazine (August, 1798).

6. If you were given sweeping powers to reduce water pollution in our society, what steps would you consider?

SUGGESTED READINGS

Most of the engineering texts discussed in the next chapter have sections describing water pollution. Other sources covering the biological impacts of pollution include:

1. H. Heukelekian and N. C. Dondero (eds.), Principles and Applications in Aquatic Microbiology (New York: John Wiley & Sons, Inc., 1964).

2. Lowell E. Kemp, W. M. Ingram, and K. M. Mackenthun, Biology of Water Pollution (Cincinnati: U.S. Department of the Interior, FWPCA, 1967).

3. Charles G. Wilber, The Biological Aspects of Water Pollution (Springfield, Del.: Charles C. Thomas, 1969).

Management approaches to water quality are discussed in:

4. Allen V. Kneese and B. T. Bower, Managing Water Quality: Economics, Technology, Institutions (Baltimore: Johns Hopkins Press, 1968).

5. National Association of Counties/Research Foundation, Community Action Program for Water Pollution Control (Washington, D.C.: 1967).

For measurement of water pollution, the accepted authority is:

6. Standard Methods for the Examination of Water and Wastewater, 12th edition (New York: American Public Health Association, 1965).

For a relatively comprehensive discussion of eutrophication, see:

7. National Academy of Sciences, Eutrophication: Causes, Consequences, Correctives (Washington, D.C.: National Academy of Sciences, 1970).

Chapter 5

WATER COLLECTION AND TREATMENT

> "We believe all citizens have
> an inherent right to the enjoy-
> ment of pure and uncontaminated
> air and water and soil; that
> this right should be regarded
> as belonging to the whole com-
> munity; that no one should be
> allowed to trespass upon it by
> his carelessness or his avarice
> or even his ignorance." Reso-
> lution, Massachusetts Board of
> Health (1869)

The Urban Water System

The Imperfect Cycle

The typical urban water-use system today is an imperfect cycle. We pump water from a local source, treat it,
use it, probably treat it again, and then discharge it back
into the river or lake to be pumped out by the next user.
But the water we return rarely has the same qualities as the
receiving waters--or the original water as it was drawn from
nature. Salts, organic matter, heat, and the other resid-
uals that characterize water pollution are left even by the
standard secondary treatment given waste water by most major
United States cities today. Aquatic ecosystems are inevitably altered, and wastes must be removed by the next user.

From the viewpoint of national policy, this circum-
stance does not inherently require change. The river, lake,
or estuary possesses natural diluting and cleansing capabilities, and, if the polluting discharge is relatively small,
the pollution effects may be negligible. Aquatic ecosystems
dependent on the sewage may even be considered acceptable or
desirable. Yet continual degradation of our natural waters
eventually invites ecological disasters, aesthetic nuisances,
and--in some cases--direct health hazards. This is particularly true of many chemical effluents from industries.

Therefore, improvement of the typical urban water system has become a national objective.

Elements of a Solution

No simple "best" strategy for improvement exists. Some water use is superfluous and can be limited or ended. Some water can be efficiently and safely recycled, especially for industries. Poor designs and inadequate maintenance for existing collection and treatment facilities often aggravate the problems and can be corrected. In still other cases, more sensitive and intelligent selection of discharge points into the natural environment can reduce ecological impact. But any solution will have to include the construction and proper operation of new waste collection and treatment facilities appropriate to the emerging pollution problems.

In the management process, these technological elements must be blended with consideration of local factors. The social, economic, and political willingness of communities to accept and finance different types of solutions has to be weighed. The characteristics of local water use have to be studied. And the nature of both the existing and potential aquatic ecosystem has to be determined. Tolerance of the biota to the final effluents will vary, and the cleansing or dilution capabilities of the natural water system will have average and seasonal characteristics.

Water Use

Average daily use of water in a typical urban system is normally considered to be about 150 gallons per capita. Only roughly sixty gallons is assumed to be for residential use, the remainder going to industrial, commercial, and government uses. For the domestic use, the U.S. Geological Survey has estimated that 41% flushes toilets, 37% is used in washing and bathing, 6% in the kitchen, 5% for drinking, 4% for washing clothes, 3% for general household cleaning, 3% for watering lawns and gardens, and 1% for washing the family cars.* (Why do we install toilets that use so much water when less greedy designs exist?)

---

*C. N. Dufor and E. Becker, Public Water Supplies of the 100 Largest Cities in the United States, U.S. Geological Survey, Water Supply Paper 1812 (1964), p. 35.

Table 5-1. Typical Municipal Water Consumption

| User | Quantity (gal./day/capita) | |
	Usual Range	Average
Domestic	20 - 100	55
Industrial	20 - 75	50
Commercial	5 - 100	20
Public	5 - 25	10
Losses	5 - 25	15
Total	60 - 300	150

In actual practice, community consumption can vary from about 60 gpcd (gallons per capita/day) to over 300, depending upon the local (1) culture, (2) climate, (3) extent and nature of industrial and commercial development, (4) price of water, (5) presence of meters, (6) water quality, (7) distribution-system pressure, and (8) the distribution-system maintenance. Some industries will draw part of their supply directly from natural sources. Some industries use more than others. Dry climates encourage lawn sprinkling, but high water costs and strict metering discourage it. Poor water quality will minimize use; and it may even drive people to competing water systems, such as commercially sold bottled water. Some leakage at joints and valves of water supply systems always occurs, but this becomes higher when (a) maintenance is neglected, (b) pressures are raised significantly above the commonly used standard of 60 psig (pounds per square inch gauge), or (c) a lack of metering complicates tracing of leakage sources.

Variations in Use

In planning or evaluating a water system, however, estimated maximum flows are normally used to select appropriate pipe diameters and storage reservoir capacities. Variations occur hourly, daily, and seasonally. Some plant operators have even claimed that sharp increases can be measured during the commercials of popular television programs when a rush to toilets occurs. During the day, hourly demand normally varies from two to seven times the average daily flow, though the average rise in demand is frequently assumed to be 4.5 times average flow. The controlling factor for pipe diameters, however, is often the water necessary for fire-fighting requirements, not the adequacy of flow to residences.

Water Supply Treatment

Sources

Local conditions dictate water sources. In cases of extreme fresh water scarcity, desalinization of sea water or, more commonly, brackish water may occur, but energy, maintenance, and material costs are high. Cisterns fed by roof-collection systems and dependent upon weather patterns and storage space are popular on islands or in coastal areas. But most urban water supplies are drawn from either surface-water sources--lakes and rivers--or ground water. Ground water can be the preferred source because (1) wells can often be drilled near the user, thus minimizing the cost of piping and pumping, and (2) the water may be relatively pure because of filtration by the soil. But neither of these conditions exist in all cases. In addition, ground water frequently has dissolved minerals that make it "hard" or objectionable because of taste or other characteristics. And the ground water supply is limited to the inflow occurring through the aquifer (water-bearing stratum), so excessive pumping and a falling water table is a chronic problem in many areas.

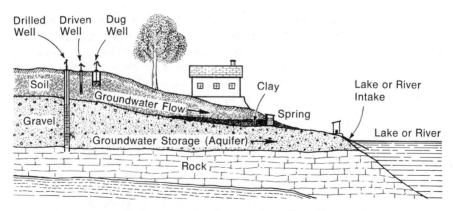

Figure 5-1. Typical sources for domestic water supplies.

Treatment Process

The required treatment depends almost entirely upon the quality of the water. The Public Health Service has published "criteria" (standards by usual definition) listing over fifty physical, microbiological, chemical, and radioactive constituents or characteristics that should be considered in designing a water-treatment process. For example, 0.05 mg/l of lead is permissible, but an absence of lead is desirable. In addition, five classes of water quality by concentration of coliform bacteria have been defined, and a treatment is prescribed for each. This may range from no treatment for protected ground water to prolonged storage followed by both special and standard treatment for severely polluted water.

In late 1974, Congress passed the Safe Drinking Water Act of 1974 authorizing the U.S. Environmental Protection Agency to set minimum purity standards that would be mandatory, which the earlier ones were not except when the water was used by interstate carriers. Standards are to be enforced by the states, but the Environmental Protection Agency can become involved if states are negligent. After February, 1977, citizens are guaranteed the right to sue offending public water systems.

Most communities depend on a relatively standard treatment process that has been refined over the past 150 years. The steps are: storage, coagulation, sedimentation, filtration, and chlorination.

Storage. Simply storing water in a reservoir, lake, or basin purifies water without the use of chemicals. Suspended solids, hardness, color, and bacteria levels fall because of temperature, sunlight, and the lack of motion. Almost all bacteria can be eliminated by storing water from one to five weeks, depending upon the temperature and chemical condition. But, since eutrophication or other undesirable effects may also occur, a retention period of two or three days is more frequently used.

Coagulation and Settling. Among the chemicals or materials added to water to precipitate undesirable minerals, chemicals, microorganisms, and suspended matter are alum (aluminum sulphate), sodium aluminate, ferrous sulphate, ferric sulphate, ferric chloride, pulverized limestone, activated charcoal, and clays. But alum or one of its chemical variants is the most popular.

When the coagulant is added by an established procedure, a flocculant precipitate--or "floc"--resembling an off-color, shredded gelatin begins to settle to the bottom of the tank. Clear water is drawn from the top and sent to the filtration units.

Filtration. Filters are usually tanks with sand bottoms underlaid with perforated pipes. As the water flows through the sand into the pipes, any remaining suspended material, including many pathogenic bacteria, are left in the top layers. Then the sand has to be either cleaned or replaced periodically.

Chlorination. Chlorination is the final, sensitive, and, for health reasons, most critical of the common treatment steps. Either chlorine gas or a liquid chlorine chemical can be used. But chlorine is difficult to handle, dangerous for operators, corrodes equipment, requires constant monitoring, and offers many other managerial problems. The chlorine dosage has to be adjusted to the water's appearance, microorganisms, pH, temperature, and the degree of treatment previously received. From ten to sixty minutes contact time between free chlorine and the water is usually considered adequate, but four hours or more may be needed to destroy a virus, such as infectious hepatitis. Destruction of Coxsackie virus has been shown to require from seven to forty-six times as much free chlorine as E. coli bacteria. Excessive chlorine has to be avoided, though, because it can combine with other chemicals in the water to form hazardous compounds.

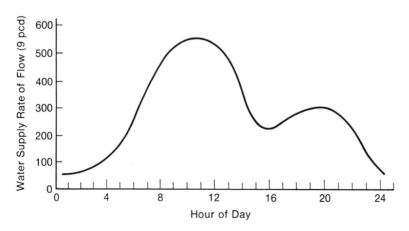

Figure 5-2. Typical hourly flow variations.

89

Relationship to Sewage Flows

Between 60 and 70% of the water entering a community
by its water supply is generally assumed to return through
sewers, the remaining 40 to 30% being lost in system leakage
or by uses that remove water from the system. This loss,
however, is typically more than replaced by leakage into
the sewerage system by ground water and rain water. If the
water table is sufficiently high and pipe joints are poorly
sealed or pipes are cracked, a constant flow of water will
be pouring into the system. During rains, storm water is
added through faulty manholes, joints, or illegal connections of residential roof drains to the sewerage pipes. It
is not unusual to find peak storm flows as high as 500 gpcd
for domestic sewers in some cities.

Waste Water Collection Systems

Design Problems

Sewerage is the system of pipes and channels that remove sewage (waste water) from residences, public buildings,
stores, offices, and factories to a treatment plant and,
eventually, points of release into the natural environment.
The overall concept of pipes carrying waste water from place
of use to place of discharge is deceptively simple. In
practice, an engineer must place the sewer sufficiently deep
in the ground that it can drain the lowest point of adjoining cellars, which are often about eight feet below the
ground surface. Yet, if pumps are to be avoided, the sewer
must slope from the places of collection with a sufficient
angle that grit and heavier wastes, such as bits of egg
shell and sand, will not collect and dam the flow. Thus,
sewage should have a velocity of more than 2.5 fps (feet per
second) but not over 8 to 10 fps or the rapidly moving grit
will wear away the pipes. Sewers should also be laid sufficiently deep to prevent crushing by traffic (about two feet)
and below the frost line (four to six feet in eastern United
States). Yet every foot of depth that must be added to the
sewer's burial means added costs, especially in areas
studded by rocks and other obstacles.

Cities must also worry about maintenance. Sewer pipes
less than eight inches in diameter are rarely installed by
municipal authorities because of the greater danger of
blockage. Also, manholes are placed periodically along a

sewer's length and at points of sudden directional change so
the sewer can be cleaned, inspected, and repaired.

Origin of Combined Sewer Systems

Historically, storm sewers preceded sanitary sewers in
the major cities of Europe and the United States. Disposal
of household waste water was considered a private matter for
back yards and an inappropriate responsibility for public
works. Connections to discharge waste waters to storm
sewers were illegal, and the waste waters were usually
poured down dry wells or cess pools. After all, these were
the days of the outdoor privies, and the tin or copper
bathing tubs were usually emptied into the kitchen sinks.
Clothes were washed by hand, and the incentive for liberal
water use did not exist.

The situation began to become more critical after the
middle of the nineteenth century. Public water supplies had
been constructed, and the quantities of water entering homes
had increased drastically. Bathrooms had been installed.
Faced with an increasingly unsanitary and unaesthetic crisis,
city administrators relented and allowed the sanitary connections to storm sewers, creating a combined system. The
concept of a combined system was extended later by land
developers trying to minimize urban development costs, but
the combined systems had generally become forbidden again
in major cities by the turn of the century.

Problems of Combined Systems

Laying one sewer system instead of two is obviously
simpler and less expensive, but this advantage is largely
illusionary. The technical requirements for the two types
of flows, sanitary and storm, are drastically different; and
mixing the two types of water has serious negative implications for environmental quality. Sanitary flows are almost
constant but relatively small in volume. Storm flows occur
only when it rains and then may exceed sanitary flows by
fifty to one hundred times. Thus, sanitary flows represent
a small trickle in the bottom of a large combined sewer
pipe; and, if the sewer is round, the sanitary flow will
spread, become slow, and solids will collect. In fact,
diameters of combined sewers may be considerably larger
than necessary for separate systems because engineers will
want to avoid any possibility of sanitary wastes backing

into cellars through storm sewer outlets, a situation that would not occur in a storm sewer alone. Also, storm sewers drain surface streets, not submerged cellars; and they do not have to be buried as deeply as sanitary sewers with all the accompanying costs. Furthermore, storm flows need a higher velocity to move gravel and other debris that is washed by storm waters, so a velocity exceeding 5 fps is usually sought.

Finally, there is the inadequacy of current technologies for separating sanitary from storm wastes. In the combined system, sewers are typically crossed by an interceptor sewer that is supposed to divert the sanitary flow to treatment plants but allows storm flows to pass unimpeded since the treatment plants could not hope to cope with the combined flow. But these interceptor sewers do not work efficiently, have a high maintenance cost, and are in themselves expensive. Since considerable solids settle in the best combined sewers, the storm waters flush this debris with its high BOD into lakes and streams. Also, storm water itself has some BOD from animal droppings, rotting vegetation, and the usual dirt and dust of city streets.

For large, older cities, such as New York and Chicago, that are burdened with combined systems and the astronomical costs of converting to separate systems, several new concepts are being tested. Usually they involve rapid treatment or the storage of the total storm overflow, either in huge rubber tanks or underground tunnels, until the treatment plant has the capacity to handle it. Then pumps send it through treatment.

Design of Storm Sewer Systems

Design of storm sewer systems should be viewed as an art, not merely a science. Certain rules have to be followed. Maximum flows are usually calculated by estimating the run-off that can be anticipated from a storm of a particular frequency, usually five years, and the changes that the run-off characteristic will undergo when the land is developed. Beyond this, the designer needs a feeling for the organic form of the land, its slopes, and its physical obstacles. Storm drainage should be incorporated into green space planning, not treated as an isolated problem that can always adequately be solved by simple engineering. Too often in the past, we have used natural streams as storm sewers. As run-offs have increased, the streams have become

unruly during floods, and engineers have been tempted to
"correct" their behavior by treating them as sewers. As
brooks have been gradually boxed on three sides and then
covered, they have disappeared from the urban scene with
corresponding losses of scenic breaks in the asphalt plains.

Waste-Water Treatment

Nature of Sewage

The purpose of sewage treatment is to wring man's
wastes from the water he has soiled. The problem is complicated by the vast amounts of water, since less than 0.1%
of the sewers' flow consists of distinct wastes, such as
grit, garbage, fecal matter, paper, rags, match sticks,
soap, fats, oil, and various complex organic and inorganic
chemical compounds. For every person in a city, the sewage
will generally contain 0.2 lbs. of BOD and 0.23 lbs. of
suspended solids daily in about 135 gallons of water.

This separation of wastes from water is not a matter
of simply constructing a standard separation process, then
leaving it to operate unmonitored. The sewage flow is always fluctuating. Composition varies even more drastically,
suspended solids and BOD being almost nil in the early
morning but rising to twice the average--or more--during
the daytime peaks. In some industrial cities, the treatment
process is complicated by "slugs" of industrial wastes that
are dumped into the sewers and eventually poison the biological treatment processes.

The Treatment Process

Sewage treatment is traditionally broken into three
stages: primary (or mechanical) treatment, secondary (or
biological) treatment, and advanced (or tertiary) treatment.
The first stage uses a few simple mechanical devices and
techniques to remove the larger and heavier bits of suspended matter. About a third of the BOD is typically removed. In the second stage, the BOD-causing pollutants are
attacked more directly by encouraging microbes (decomposers)
to break down the unwanted organic matter. In an effective
plant, sewage can emerge from secondary treatment with 10%
or less of the BOD remaining. But the nutrients and other
dissolved chemical contaminants still remain, and they must

be removed in the advanced treatment by chemicals, strainers, or other techniques. Besides these stages of treatment, the sludge--the suspended solids that are removed from the sewage at various stages--still remains and must be separately processed and returned to the natural environment.

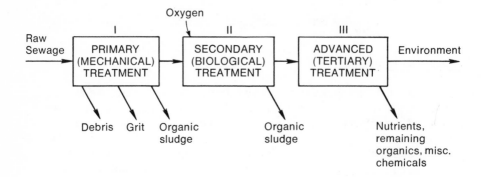

Figure 5-3. Three sewage treatment stages.

Most cities today use primary treatment, although large areas of some major cities are still outstanding exceptions. Raw sewage, for example, still pours into the Hudson River from the West Side of Manhattan. Most of the nation's sewage receives secondary treatment, and the public concern over eutrophication is rapidly causing addition of advanced treatment, especially for new treatment plants. Rural areas and outlying suburbia are still generally depending upon the septic tank, a primitive but potentially effective device that is being grossly over-used and misapplied.

Primary Treatment

Three devices are typically used for primary treatment: debris remover, grit chamber, and the plain sedimentation tank. Unless secondary treatment facilities are added, a separate sludge-storage tank must also be available. Debris --rags, large pieces of paper, dead animals, small branches, and similar bulky items--are caught by a grate or screen and removed as the sewage enters the treatment plant. Sometimes a communitor (a shredder) or similar device will grind the waste and send it into the settling tanks beyond. Even when fine screens are used and the materials are removed, however, BOD would not be decreased more than about 10%. So most plant designers are content with some simple, inexpensive devices.

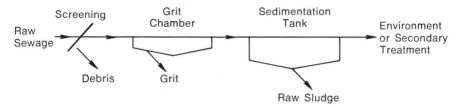

Figure 5-4. Primary (mechanical) treatment.

The sewage next passes through the grit chamber, usually a long, rectangular, box-like tank with space below the water's flow for sand, gravel, egg shells, bits of bone, and other small, heavy particles to settle. Usually an automatic scraping device will pull the grit to one side or into a corner where it can be removed. By increasing or decreasing the flow of water, thus the water's velocity, a plant operator can usually select or reject finer particles.

Finally the sewage enters the plain sedimentation tank, called "plain" because no chemicals are added. The tank may be rectangular or round, but skimmers must be able to travel across the top of the water to remove grease while scrapers or rakes travel across the bottom to put the sludge into a drain for removal to a sludge tank. Sewage may remain in the tank from forty-five minutes to two hours, more than two one-half hours being considered the maximum worthwhile period. Between 50 and 60% of the suspended solids can be removed in plain sedimentation, and the BOD content generally drops by 30 to 50%.

Secondary Treatment

The secondary treatment unit typically consists of two tanks. But to this is usually added a sludge digesting unit. Most secondary treatment facilities are classified as activated sludge or trickling filter.

Activated Sludge. In the activated sludge process, the sewage is first "seeded" with microbe-laden sludge, and the mixture is circulated in a tank with bubbled air for four to eight hours, six hours being the common standard. (Circulation time can be drastically reduced by substituting oxygen

for air.) The sewage could be aerated without the prior
addition of "seed" sludge, but the process would be considerably less efficient. In effect, the treated sludge contains an immensely complex ecosystem rich in decomposers.
Specialized unicellular bacteria and algae, numerous protozoa, and some metazoa thrive in the proportions that can
rapidly oxidize the raw sewage's organic material.

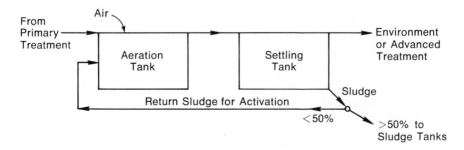

Figure 5-5. Activated sludge process for
secondary (biological) treatment.

After the aeration tank, the sewage passes into the
final settling tank where scrapers remove the sludge; and,
after some one to two hours, the liquid is released into an
advanced treatment unit or the natural environment. Some of
the sludge is returned to the aeration chamber to activate
raw sewage while the remainder goes to holding tanks or a
digestion unit. In the activated sludge process, suspended
solids are usually decreased by 85 to 95%; BOD, 80 to 95%;
and coliforms, 90 to 95%. The primary disadvantage is the
sensitivity of the sludge to varying water conditions, especially in an industrial area where industrial chemicals may
poison the decomposers and halt oxidation for hours or days.
Meanwhile, raw sewage will continue to flow into the plant.

Trickling Filter. One common alternative is the trickling filter, which consists of large beds of crushed stone,
gravel, or slag which are doused either periodically or continually with sewage. Sometimes a circular bed is used with
the sewage distributed by a rotary boom that is often 100
feet in diameter. A less commonly used alternative is a
rectangular bed with fountains that spout periodically.

In either case, stones are covered with a transparent
film that, like activated sludge, contains a complex ecosystem of microbes. These remove suspended solids and gradually break down the constituents into basic compounds. As in

96

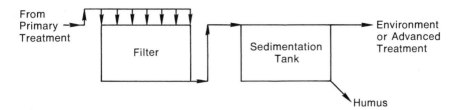

Figure 5-6. Trickling filter process for
secondary (biological) treatment.

most alternatives, though, the effluent must be passed
through a final sedimentation tank to remove the humus,
the dead organic matter from filter stones, and any final
particles that may have escaped the filter. Except in unusually cold weather, the efficiency of trickling filters is
approximately the same as activated sludge. However, health
authorities are becoming wary of dangers posed by the aero-
sols formed when sewage flows from the boom or splatters on
the rocks. Furthermore, trickling filters do not have the
flexibility of flow possessed by activated sludge systems,
and they can also be sensitive to industrial effluents.

Sludge Disposal

     Sludge characteristics can vary significantly from dif-
ferent types and levels of treatment. Most sludges, though,
are more than 95% water, have offensive odors, and will not
dry readily if spread on beds. To make the material more
innocuous, the sludge is normally digested--broken into
simple compounds that are either useful or not offensive.

     If kept in a tank, sludge will naturally pass through
three phases. Initially, the dense liquid will become highly acidic, carbon dioxide and nitrogen will be released in
large quantities, and the odors from sulphur dioxide will
increase. The BOD levels will rise. In the second phase,
the acidity will diminish, although hydrogen sulphide, carbon dioxide, nitrogen, and some hydrogen will still be released. Finally, in the third phase intensive digestion
will occur. Methane gas will be emitted profusely with some
carbon dioxide and nitrogen. The BOD will rapidly fall.

     Left to nature, this evolution will take months, but
it can be speeded by biological engineering. In a sludge
digester, the temperature is held to about 100° F. and an
alkali pH is maintained. Fresh sludge is added slowly, not

97

exceeding 2 to 4% by weight of the digestive solids already in the tank. Methane is drained off and usually used for heating, between 600 to 700 BTU being contained in the 0.7 cubic feet of methane produced per capita daily. The digested sludge can be easily dried on drying beds or in a vacuum filter. When dry, it can be incinerated or used as landfill.

In a patented process that has been used by some cities, the sludge has been reduced to ashes by a wet oxidation process. The sludge is injected into a reaction chamber where air pressures are between 1,200 and 1,800 psig and the temperatures are 540° F. or more. Heat is removed from the reaction by circulating the liquid through heat exchangers. In at least one case, Chicago, the process has been abandoned because of the expenses associated with skilled operators, high corrosion levels, and the elaborate equipment. (Chicago is now using sludge for land recovery.)

Advanced Treatment

Advanced or tertiary treatment is still sufficiently new that no standard methods have clearly emerged. In some cases, chemical treatment with sedimentation is being used. In others, a rapid sand filter or even mechanical filter is being favored. A combination of approaches is being preferred by still others. But the common objective is to remove 98 to 99% of the BOD and other potential pollutants so the water can be returned to nature in essentially a pure state.

Septic Tanks

The septic tank is a safe and adequate means of sewage treatment--if the tank and the drainage field are properly constructed in appropriate soils and provided sufficient maintenance. If the rules are followed, a septic tank consists of a concrete or steel container of sufficient size to hold sewage under anaerobic conditions for at least twenty-four hours, then allow the effluent to pass into a drainage field where the liquid can filter from perforated pipes into the soil. Periodically, the undigested solids are pumped from the tank and removed to be treated as sludge.

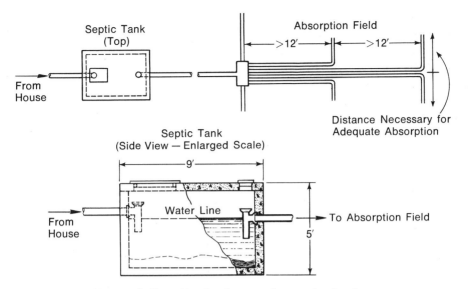

Figure 5-7. Typical septic tank design.

In practice, the much-maligned septic tank is typically
so overloaded with sewage that materials simply pour in one
side and out the other. If the soil is insufficiently por-
ous, the raw effluent rises to the soil surface and dribbles
into the nearest stream. And householders will rarely go to
the expense and trouble of having their tank uncovered and
pumped out, so the drainage field becomes clogged and solids
either emerge from the ground or sewage backs into the house.
Some sewage in streams could even be accepted if the residences were isolated farm houses, and the streams could
sufficiently dilute the contaminants. But this is not
possible in the suburban housing developments where square
miles are studded with houses, each with a septic tank that
has been inadequately constructed by unscrupulous developers.
As one saying goes, "The septic tank is like an honest country cousin who was taken to the city and went bad."

Package Plants

One recently developed answer to the problem of isolated developments is the package sewage plant. These are
factory-built units that can be transported to the site and
bolted quickly together into an operating treatment plant.
In effect, they are miniaturized activated sludge or trickling filter units with built-in sludge digestion. To be
semiautomatic, capacities are generally larger than

specified; and the power requirements are relatively high. However, where a development has under several thousand homes, and the city's sewers have not arrived but treatment is necessary, package plants fulfill a necessary function.

Lagoons

Lagoons, the open ponds where sewage is collected and allowed to react naturally, represent one of the earliest forms of sewage treatment that have earned a well-deserved reputation for foul odors and wasted land. In recent years, though, new equipment has been designed to provide aeration for the lagoons. With seeding from older lagoons, the modern aerated lagoon resembles an activated sludge process. Where temperatures are consistently high and land values are low, aerated lagoons may have a distinct advantage over alternative methods of treatment.

Other Treatment Processes

The concept of adequately treating sewage with two stages, primary (mechanical) and secondary (biological)-- and possibly a third, advanced--has dominated the field of sanitary engineering for decades with relatively few innovations. A few variants of the activated sludge or trickling filter processes have been introduced, but the fundamental concept of passing sewage over or through a biologically active medium has remained and can be found in virtually every major treatment plant operating today.

However, economics, stricter regulation, and innovative concepts have been moving new plant designs into less familiar patterns. Physical-chemical treatment, which omits the biological stage and provides a treatment process somewhat resembling that given to water supplies, has been introduced in some cities. Supplemental use of chemicals to improve precipitation at the first or second sedimentation tank is increasing. Filtration by passing sewage through either microfilters or beds of coal or sand is being tried. Adsorption, the removal of dissolved organic materials by passing sewage over beds of activated charcoal, is being advocated by some authorities, especially when industrial pollutants are present. Dialysis, the use of semipermeable membranes to filter out some pollutants, has been explored; but the high cost and unsatisfactory efficiency of existing membranes have restricted use to a few industrial applications. And there have been experiments with rows of large discs that look like wheels with only part of the rim submerged but serve the same treatment function as the stones in a trickling filter.

In terms of capital expenditures, the funds being spent by industry for construction of waste water treatment plants are estimated to exceed those of municipalities by a ratio of five to one since the Federal Water Pollution Control Act Amendments of 1972 set the years 1977 and 1983 as compliance dates. Treatment processes used are generally the same as those used by municipalities, though, except that the concepts can often be simpler because the waste water characteristics are usually more consistent and less complex. In some cases, such as parts of the paper industry, industrial processes are being changed to provide an effluent that is easier to treat.

Operational Faults

Improper design, supervision, and operation of waste water treatment plants reduce their effectiveness and represent a widespread, chronic problem. In the design stage, each plant is normally custom-engineered. Some operate according to expectations, others do not. More than two-thirds of all municipal plants reportedly fail to meet Public Health Service standards for personnel qualifications, laboratory controls, and maintenance of records. Most have operational, mechanical, or structural problems; and a significant number regularly bypass some sludge.

Bypassing sludge means the flushing of sludge directly into a stream with the treated water. This negates gains in the treatment, but it also diminishes the costly and messy difficulties of hauling and disposing of the sludge. Since sludge settles quickly to the bottom of a stream, bypassing can be difficult to detect, especially when occurring at night. While bypassing sludge can represent a form of petty graft or an expression of laziness, it more commonly is used as a means of meeting inadequate budgets and is a logical response to excessive penny-pinching by a city council.

Water-Quality Planning and Management

Legal Framework

Federal water-quality legislation today has two roots, one in a series of progressively stronger Congressional acts passed since 1948 and another in the long-forgotten Section 407--better known as the Refuse Act--of the Rivers and Harbors Act of 1899, that was brought to public attention in the late 1960s. The 1948 law was passed during a period of intense consciousness of states' rights, and the weak provisions for federal water-pollution research and urban sewer grants were possible only because of a loosening in the intense decades-old opposition by congressmen representing coal-mining and industrial interests in the Ohio Valley. The Water Pollution Control Act of 1948 required the U.S. Public Health Service to provide technical information to the states and established a Water Pollution Control Division within the Public Health Service. This division was never active, and the small amount of money available for building sewers declined from an appropriation of $3,000,000 in 1950 to less than $1,000,000 in 1955.

Federal intervention resumed with the Water Pollution Control Act of 1956, which funded additional research into pollution problems, provided matching grants for urban sewage-treatment plants, and stipulated a conference procedure for combating interstate water pollution. Because of opposition by President Eisenhower to federal involvement in pollution problems, the conference mechanism was not used and sewer-grant funds were vetoed. Amendments to the Federal Water Pollution Control Act in 1961 were notable only for authorizing seven new research laboratories in politically sensitive congressional districts.

With water-pollution problems obviously intensifying, the federal role was strengthened significantly by the Water Quality Act of 1965. Large appropriations for construction grants and a transfer of water-pollution control from the Public Health Service were provided. But the key provisions were for states to prepare stream-quality standards by June 30, 1967, and to provide implementation plans for meeting the standards. In the following year, the Clean Water Restoration Act of 1966 expanded the grant program with funds and more liberal provisions. The need for regional, systematic planning was recognized but not supported with adequate funds.

By 1970, the futility of the 1965 approach was becoming obvious. State standards tended to be weak, implementation schedules were either falling behind or were nonexistent, and federal appropriations had shrunk under the financial pressures of the Viet Nam war. The response by Congress was to pass the Water Quality Improvement Act of 1970 that provided stricter control of oil pollution, wastes from water craft, and other special cases. But grants for construction remained the primary approach to improving water quality, and relatively few changes were made in planning or enforcement mechanisms. Pollution from industries and from cities not receiving grants generally continued to increase.

Against this background, the Rivers and Harbors Act of 1899 was activated. Sections 407, 411, and 413 forbade the dumping of any refuse into navigable waters or tributaries without permits and provided for enforcement, but the Act exempted municipal sewage. The program to issue permits never succeeded. Processing of applications provided an administrative nightmare; litigation developed over the need for environmental impact statements and other legal interpretations. Some industries were quick to take advantage of the exemption for municipal sewage by diverting-- legally or illegally--toxic industrial wastes into municipal systems. Efficiencies in the secondary treatment processes fell, and pollution levels in streams rose.

Stung by failures of the past legislation and annoyance among voters, Congress passed the Federal Water Pollution Control Act Amendments of 1972. This exceptionally lengthy, complex legislation provided an elaborate, ten-year schedule for establishing national water-quality standards and steps in implementation. But the key enforcement mechanism was a shift of emphasis from water-quality standards and conferences to effluent limitations and prompt court action.

Although the concept of stream standards was strengthened by granting increased standard-setting authority to the federal government, key provisions were the establishment of July 1, 1977, as the deadline for industries to install the "best practicable control technology currently available" for point sources discharging into navigable waters and July 1, 1983, for installation of the "best available

treatment economically achievable."* The latter refers to
levels of treatment attained by the most effective equipment
found in a particular industry--after government specialists
have added refinements. Self-monitoring of discharges by
industries is required. Also, the law provided the federal
government authority to delegate to states the responsibility for issuing permits required by the Refuse Act of 1899,
and the permit requirement was expanded to encompass municipal and other point sources. Over 40,000 permits are estimated to be involved in the resulting National Pollutant
Discharge Elimination System (NPDES). Maximum duration for
a permit is five years.

The Act also provided matching funds for river basin
planning and spelled out numerous deadlines for standards
and plans. Provisions were made for several special water
quality problems, such as thermal discharges, oil pollution,
and sludge disposals. One section exempted some EPA actions
from NEPA's impact-statement requirements, and another permitted some types of private suits.

Difficulties arose immediately. President Nixon, who
initially vetoed the legislation, indicated he did not
intend to spend more than a third of the authorized $18.35
billion. Industries expressed uneasiness over the requirement to provide cost and production-related data to the
government, and the law's inflexibility provoked widespread
complaints. Yet the law introduced new control concepts
and has been unquestionably more successful than any pre-
ceding legislation.

Regaining a Heritage

Water represents one of the basic human needs, and in
these days of widespread water pollution, we easily forget
the simple joys of a clean environment. But we sense a
world that could be in the descriptive fragments written by
early observers of rivers familiar to most of us. In 1524,
Giovanni de Verrazano entered the Hudson River and was en-
chanted that "the trees exhaled the sweetest odors." Even

---

*Do not be deceived by the seemingly restrictive terms.
"Point sources" has been defined as any pipe, ditch, channel, tunnel, vessel, feedlot, vehicle, or anything else that
could convey waste water. "Navigable waters" was defined in
the act as meaning "waters of the United States."

in the nineteenth century, a resident of Albany could still describe the Hudson River as "clear as crystal and fresh as milk."

QUESTIONS FOR DISCUSSION

1. Describe specific situations in which you think one of the three typical waste-water stages may be justifiably used alone.

2. Explain how you would justify addition of a tertiary (advanced) sewage treatment stage to an existing primary-secondary sewage treatment plant in a Midwest city.

3. Assuming that you were appointed commissioner of water resources in your city, list three actions that you would consider taking and briefly justify selection of each.

4. If the most advanced waste-water treatment techniques were provided, would you advocate recycling your city's sewage as water supply? Explain your answer.

5. Describe steps that could be taken in a typical suburban household to diminish water use.

SUGGESTED READINGS

Numerous excellent references for water-quality control have been written but, except for the type mentioned in the previous chapter, all are intended for engineers.

1. G. M. Fair, J. C. Geyer, and D. A. Okun, Elements of Water Supply and Wastewater Disposal (New York: John Wiley & Sons, Inc., 1971). Considered the classic in sanitary engineering, but is highly technical.

2. W. Wesley Eckenfelder, Jr., Water Quality Engineering for Practicing Engineers (New York: Barnes & Noble, 1970). A concise, low-cost handbook on waste-water engineering. Only for the technically inclined.

3. American Water Works Association, Water Quality and Treatment: A Handbook of Public Water Supplies, 3rd edition (New York: McGraw-Hill Book Company, 1971). The standard reference for water-supply treatment.

4. James F. Johnson, <u>Renovated Waste Water: An Alternative Source of Municipal Water Supply in the United States</u> (Chicago: University of Chicago, 1971). Available for $4.50 from the Department of Geography, University of Chicago, 5828 South University Avenue, Chicago, Illinois 60637. Specify Research Paper No. 135. Funded by Resources for the Future. Frank approach to a situation closer to reality than commonly thought.

ESSAY

The Great Water Crisis of 1982

Causes and Effects

We will have a Great Water Crisis. The year may not be 1982, but the event's appearance within the next decade or two seems as inevitable as the movement of the planets. The immediate cause will be a drought in the East or Midwest. As one or two years of reduced rainfall cause water levels in storage reservoirs to drop, ominous predictions of short-ages will begin to appear in newspapers and political fig-ures will plead for water conservation. Perceptive observ-ers will explain the difficulties by pointing to three circumstances: (a) a rising use of water in recent decades, (b) an earlier reluctance by cities to construct treatment plants appropriate for the available new sources of water, and (c) the spread of cross connections in century-old water systems.

Because of unusually wet weather conditions during the late 1960s and early 1970s, plants to treat new sources of water supply were not developed at the same pace as rising use, which almost doubled over a decade in some cities. Thus, cities faced with inevitable drought conditions must cope with the dilemma of the only available new water supply being dangerous to health; but, to prevent contamination of the system from cross connections, pressure must be main-tained by pumping more water into the system as quickly as it is used or leaks out.

Questionable Sources

Since their water systems were organized, many cities in the Northeast, such as New York, have drawn water sup-plies solely from wells or reservoir areas void of signifi-cant industrial or domestic pollution. Avoidance of major lakes and rivers, such as the Hudson River, has become a political creed, especially after publicity during the past decade has created an image of these waters being saturated with pesticides, heavy metals (especially cadmium, mercury, lead, and arsenic), and biorefractories, the chemical com-pounds created during industrial processes and including some known carcinogens (cancer-causing compounds). In late

107

1974, for example, newspapers throughout the country reported that a study had identified sixty-six chemicals,
mostly biorefractories and including known carcinogens, in
New Orleans' drinking water. In addition, cancer mortality
for white males using the city's water had been found to be
15% higher than for similar persons drinking well water.
(Disposal of industrial wastes by both legal and illegal use
of deep-well injection, the pumping under high pressures of
liquids into the ground, did not receive comparable notice
and may be an even more significant long-term hazard.)

Under the Water Pollution Control Act of 1972, discharge of most of these industrial chemicals will be curbed,
and the glare of public notice will undoubtedly hasten the
process. Another health threat that is not affected by the
1972 law, however, is the presence of viruses in water supplies. Considerably smaller than bacteria and, under some
circumstances, incredibly hardy, different viruses cause a
wide array of illnesses ranging from common colds and flus
to polio and infectious hepatitis. Some outbreaks of hepatitis, especially, have been traced to water-supply contamination. Yet we do not even possess a reliable means of
routine testing for the presence of viruses, and the Public
Health Service's Drinking Water Standards issued in 1962 did
not set any virus standards.

Even without tests available, plants can be constructed
to cope with viruses and, if they remain, the biorefractories and heavy metals. Construction of a new water-treatment
plant, though, can require from three to ten years, especially if the designs have to be developed. This would
obviously be a difficult way to respond to the Great Water
Crisis. Strict conservation measures, especially for industries, could reduce use; and linkages with adjoining
water systems could permit more purchases by supply-short
cities. Nevertheless, the nervousness of officials responsible for water supplies will be markedly increasing because
of the inadequacy of the pipeline system that typically
carries water to the urban consumer's faucet.

Cross Connections

Cross connections are any openings between the water
supply and other liquids, including sewage and industrial
chemicals. If the pressure in the water-supply pipes ever
falters, either by mishap or an inadequate water supply,
vacuums will develop at various points in the system and

contaminants will be drawn into the water-supply pipes. Cross connections are rarely deliberate. Plumbing inside equipment, such as an ice-making machine or a water cooler, is a frequent cause. Valves can stick in the plumbing of large, complex facilities, such as hospitals. Sometimes industries will use secondary water, water pumped from a river or lake, for use in cooling or flushing. Secondary water storage tanks will then be connected to the municipal water-supply system through, for instance, the priming device of a pump. At least one incident involved a cross connection in a university's irrigation system.

Cross connections can be expected simply because many of our urban water-supply systems are ancient, often exceeding a century in age; and they have been amended and modified countless times. Pipes can rust, become crushed, or start leaking through aging joints. Cold water entering the pipes each fall will cause pipe contractions, creating leaks in pipe joints that can be submerged in ground water or leakage from nearby sewers.

Precautions are customarily taken against cross connections. Every city employs inspectors to approve plumbing of new buildings or renovations. Building codes insist upon safety valves or air gaps, the absence of any direct connection between water supplies and other liquids. Nevertheless, the plumbing systems in downtown areas of cities defy inspection or understanding. Many modifications occur without the city being informed, and the overworked, underpaid inspection officers are loath to seek more problems.

Back to Normalcy--Almost

The Great Water Crisis will not strike all Northeastern or Midwestern cities since many are already drawing their water from low-quality, high-volume sources and are in the midst of a crisis over the presence of industrial pollutants. Nor will the West and Southwest be involved because they have long been conscious of the need to conserve and develop limited water supplies. Even for the affected communities, the crisis will eventually pass and join the Energy Crisis in history. But some styles of life will change. Water prices will rise drastically to pay for system improvements, and people will feel a tinge of guilt when a shower is run more than a few minutes or clothes are washed after only a day of use. Water consumption will become another criterion to note when buying appliances.

In one more dimension, the finite nature of the world's resources will have become obvious.

Chapter 6

COASTAL ZONE MANAGEMENT

"Beyond all things is the
ocean." Seneca (c. 50 A.D.)

Overview

The focus of the coastal zone is the estuary, the
marine area where fresh and salt water mix. Estuaries are
the bays, river mouths, and coastal lagoons where the fresh
waters of surface and subterranean streams flow into the
sea, carrying nutrients, organic materials, and sediments
to create some of the most biologically active natural zones
of our earth.[*] Directly or indirectly, the majority of advanced sea life appears dependent upon the estuary for some
aspect of its existence. Yet estuaries have also become
foci of human activity, and destruction of the estuarine
ecosystems may emerge as the most tragic and momentous of
man's environmental follies.

Physical Characteristics

By one viewpoint, the typical estuary would seem a
harsh environment for life. Everything appears in constant
transition. Daily tides sweep the waters back and forth.
Salinity--the proportion of salt in the water--changes both
horizontally and vertically as currents mix the fresh and
salt waters with their differing densities, temperatures,
and chemical properties. Bits of suspended matter carried
by the fresh water are continually dropping to the bottom,
tending to smother bottom organisms.

Superimposed upon this daily fluctuation are the seasonal variations with torrents of fresh water pouring into
the estuary during spring floods to be followed by a mere
trickle during the late summer. Because the shallow water
is exposed to the atmosphere's fickle weather patterns, temperatures can drop precipitously in winter, freezing all

---

[*]"Estuary" is assumed to derive from Latin Aestus--tide, or
boiling.

biota that cannot flee to deeper, more stable water. In the summer, the same estuarine shallows soar in temperature, and evaporation can cause salinities to rise as high as 80 ppt (parts per thousand) in some southern coastal lagoons compared to the normal ocean levels of about 35 ppt.

Yet the estuary can also be described as extraordinarily suited for the promotion of life. The mixture of fresh and salt waters creates a gentle broth that is warmed by the sun's energy to more biologically active temperatures. This broth is continuously fertilized by the organic material and nutrients washed from the vegetation and dissolving rocks of the land as well as the nutrient-rich waters of the deep ocean. And the entire mixture is effectively stirred and rocked by tides and currents. Both descriptions are true, and the result of this contradictory nature is typically a high rate of growth for all organisms but an underlying fragility that can be easily upset by significantly changing just one physical characteristic.

Mixing

Fresh water will usually differ from sea water in temperature, suspended solids, current direction and speed, and other characteristics; but the salt content is often the key factor in controlling both biological adaptation and the mixing process. In a typical estuary, salinity is considered to range from about 0.5 to 30 ppt. But even "fresh" water has some dissolved salt, most rivers in eastern United States carrying about 0.1 ppt or less compared to the approximately 35 ppt of the open sea. But the chemical compositions of the salts are different. In sea water, about half of the dissolved solids include the chloride ion (sodium chloride is common table salt) while only one-tenth to one-twentieth of fresh water's dissolved solids can be associated with chloride.[*]

Being lighter than salt water, fresh water tends to rise when it enters the estuary. A salt-water "wedge" is formed, and mixing occurs along this fresh-salt water interface. The length and shape of the wedge will depend upon both the flow of the river and the shape of the estuary. In the Hudson River, for example, the salt-water wedge reaches

_____

[*]Analysis is complicated because the chloride exists in the sea water as a free ion, not part of a complete molecule.

as far as Albany during a part of the year. Yet in a few
estuaries that empty directly into deep water, the wedge may
be almost completely in the sea.

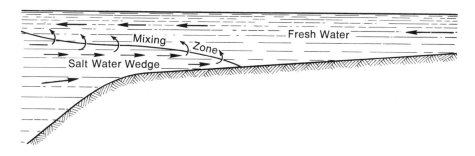

Figure 6-1. Salt-water wedge and mixing
patterns in a typical estuary.

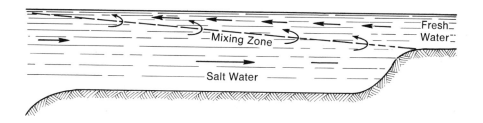

Figure 6-2. Mixing pattern when a fresh-water
stream empties directly into deep water.

Because of the rotational force of the earth, the
fresh-water stream in the northern hemisphere will tend to
move towards the right bank of an estuary. The mixing is
further complicated by the tidal movement that varies from a
few inches in the Gulf of Mexico to twenty-six feet or more
in the North Atlantic. One effect of tides is to increase
the mixing, some materials carried by fresh water being deposited in the salt water only to be returned to the fresh
water again when the tidal current moves the salt wedge upstream.

113

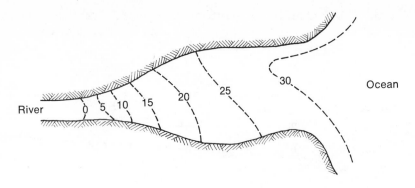

Figure 6-3. Average vertical salinities
in a typical estuary.

Sedimentation

Sediments shape the topography of an estuary and provide the bottom strata, two factors that strongly influence
the biota that the estuary will possess. Sediments origi-
nate as the fresh water's suspended and dissolved solids
that are moved down the stream bed from the drainage area,
and in the estuary they become the bottom muds, sand banks,
and gravel deposits. The proportions of each material will
depend upon the rocks, soils, vegetation, and disruptions
occurring in the drainage area, plus the carrying capacity
of the river at different stages.

The materials are finally deposited by several mechanisms. Heavy sands can be dragged along the bottom of the
river and often come to rest when they hit the tip of the
salt wedge. But since the salt wedge may be moving according to tides and the river flow, the deposition will typically be in a broad band. Fine sands carried higher in the
river's current may be deposited through the salt wedge as
the current slackens. Clays, other fine materials, and some
dissolved solids may be chemically flocculated--caused to
come together as larger masses--in the salt water and eventually be deposited as a mud. But, being light, the flocculate may be recirculated back into the fresh water by the
upstream-moving salt wedge. Once in the lower estuary, the
flocculate may drift in a different pattern than would be
expected from the dominant currents.

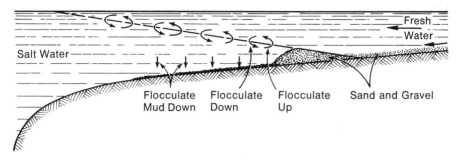

Figure 6-4. Deposition of sand and clays
in a tideless estuary.

Nutrient Levels

Estuaries are fertilized by at least four mechanisms:
(1) the leaching of plant nutrients from the soil in a
river's watershed, (2) the decay of organic materials
(detritus) that wash into the river or estuary, (3) pollution throughout the river and estuary system, and (4) the
current of nutrient-rich sea water from the lower ocean
strata that are the source of the salt wedge. Once in the
estuary, these nutrients tend to be trapped by the conflicting water flows and the precipitating muds. But even the
nutrients that escape the estuary can influence the sea surface for hundreds of miles. The important role for the
nutrients, though, is in the estuary.

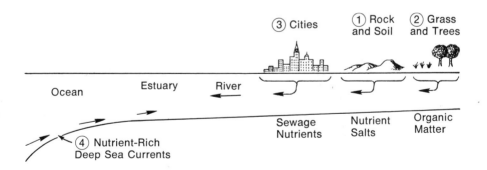

Figure 6-5. Sources of nutrients for the estuary.

While the oceans have a vast reserve of nutrients, they
are thinly scattered and are usually concentrated in the

lower levels. In the estuary, the nutrients are in the
upper layer of water where, combined with the sun's energy,
chlorophyll can be formed and plants can build organic
matter. For example, Bostwick H. Ketchum found that samples
of water (30 ppt salinity) taken from the Hudson River
estuary had about six times the concentration of phosphorus
as surface coastal water, and two and a half times as much
as deep ocean water.

Biological Characteristics

Primary Production

Despite their fragility, estuaries are consistently
among the most productive zones in the world in terms of
average plant growth. New organic material is formed by
photosynthesis at a rate of less than 1.0 gms/m^2/day in the
open ocean, less than 0.5 gms/m^2/day in deserts and semi-
arid grasslands, 0.5 to 5 gms/m^2/day in normal agriculture,
fresh-water lakes, most grasslands and average forests, but
5 to 20 gms/m^2/day in estuaries, coral reefs, evergreen
forests, and exceptionally intensive agriculture. In the
mid-fifties, Long Island Sound between Connecticut and New
York was estimated to be producing an annual average of
about 3.2 gms/m^2/day of vegetable matter; and some exceptionally productive estuaries, such as Silver Springs,
Florida, attained 17.5 gms/m^2/day. When the grass in an
estuary dies and decays in the water, some nutritional
properties, such as protein content, may rise even further

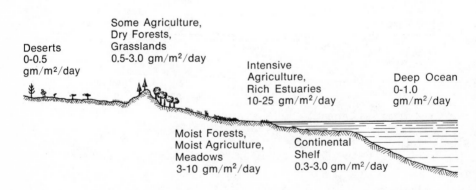

Figure 6-6.   Typical spectrum of primary production
in different biological zones.

116

because of the rich bacterial concentration. In addition, estuaries benefit from organic material, including bacteria, brought from distant terrestrial systems by the fresh-water streams.

The Consumers

With these opportunities, the existence of an unusual variety and profuse quantity of fauna in the estuary should not be surprising. Yet the estuarine ecosystem is complex and sensitive with the distribution of the fauna determined by numerous interrelated physical conditions, including currents, salinity, temperature, bottom material, calmness of the water surface, dissolved oxygen, stability of fresh-water flow, turbidity of the water, and the variations in water depth. Some estuaries are particularly rich in fauna while others are relatively poor both in quantity and variety because of either the natural conditions or alterations by man.

Ecologically, we can classify the estuarine fauna by their habitat (physical environment), niche (functional position within the habitat), and estuarine dependence (degree that estuarine habitat is essential to species perpetuation). The first two classifications illustrate in more detail the fragile complexity of the estuary, but it is the third that sets the final stage for examining the impact of man's activities upon the estuary-related biosphere.

Habitat. Several systems have been devised to classify estuarine fauna according to the estuarine zone where they are found. One system describes organisms according to their characteristic depths within the three tidal zones: supratidal (above the high-tide mark), intertidal (between high and low tides), and subtidal (below the low-tide mark). In North Carolina estuaries, researchers have contrasted the sequence of biota found on rocky bottom material with that of sand substrata. Algae and shellfish predominate on the rocks, while crustaceans are more prevalent on the sand.

This system is obviously not helpful for fish species that must remain below the water surface regardless of tidal fluctuation. So another approach to classification divides the estuary into four salinity zones--the estuary's mouth, lagoon, middle reaches, and head. For example, 310 species of fauna were identified in the particularly rich Knysna

Table 6-1. Distribution of Bottom Fauna on
(a) Rocky Shore and (b) Sandy Beach
in Beaufort, N.C., Estuary

Zone	Rocky Shore	Sandy Beach
SUPRATIDAL ZONE Mean High Tide	Periwinkles (Littorina) Encrusted black lichens (Myxophyceans) Isopods (Ligia)	Ghost crabs (Ocypode) Beach amphipods (Talorchestia, Orchestia) Tiger beetles and other land insects
INTERTIDAL ZONE Mean Low Tide	Barnacles (Chthamalua, Balanus, or Tetraclita) Oysters (Ostrea) Green algae (Enteromorpha and Ulva) Mussels (Mytilus)	Ghost shrimp (Callianassa) Burrowing isopods (Chiridotea) Burrowing amphipods (Haustorius) Polychaete worms Mole crab (Emerita) Beach clam (Donax)
SUBTIDAL ZONE	Red or brown algae (Laminarians and Fucoids) Sea anemones (Aiptasia) Sea urchin (Arbacia) Corals (Leptogorgia and Oculina)	Sand dollar (Mellita) Burrowing shrimp (Ogyris) Acidians (Thyone) Heart clam (Cardium) Olive shell (Oliva) Copepods Sea pansy (Renilla)

Source: Pearse, Humm, and Wharton (1942) and Stephenson and
Stephenson (1952).

Estuary of South Africa. The largest number, 199, were
found in the relatively saline, productive, and calm lagoon,
but the more turbulent mouth trailed closely behind with 179
species. The less nutrient-rich middle reaches (sixty-four
species) and head (thirty-three species) trailed far behind.
The overwhelming majority of these species are considered
primarily marine (oceanic).

Table 6-2. Distribution of Fauna Species
in the Knysna Estuary, Kenya

Salinity Range (ppt)	Mouth 34.5- 35.7	Lagoon 29.1- 34.8	Middle Reaches 18.9- 26.5	Head 1.1- 14.0	Total Species
Fresh water species	0	0	0	7	7
Estuarine species only	3	20	12	12	27
Estuarine-ocean species	80	108	46	14	137
Predominantly ocean species	96	71	6	0	139
Total species	179	199	64	33	310

Source: J. H. Day, "The Biology of Knysna Estuary, South
Africa," in Lauff (ed.), Estuaries.

Niche. The variety of estuarine organisms and their
seemingly simple geographic distribution over the estuary's
different zones begins to fall into a more complex logical
scheme when functions, especially in the food chain, are
examined. Plants have become specialized in different
micro-habitats. Attached plants, marsh grass, for example,
obviously cannot grow in the middle of a deep-water channel.
But other, less obvious, factors--such as tidal flow, salinities, mud characteristics, and available sunlight--also
influence plant location. And, whenever feasible, each
important type of plant life will have acquired one or more
species of fauna, usually fish, that has become particularly
adept at harvesting the plant's stored energy. This establishes the base of the food chain described in Chapter 2.
Carnivores will have evolved to prey upon the herbivores
(plant feeders), and secondary carnivores--such as man--
will prey upon the primary carnivores, including tuna and

swordfish. Because of the limits on appropriate food
sources, the higher level carnivores must be (a) versatile
in their feeding habits, (b) wide ranging in their habitat,
or (c) rare.

Estuarine Dependence. Visitors and inhabitants of
estuaries--the fish, birds, insects, animals, and other
mobile biota--possess a phenomenal variety of life cycles.
What would happen if the estuary were destroyed? A few
particularly adaptable birds, fish, and animals may scarcely
be bothered since use of the estuary appears to be a habit
or only an occasional convenience, not a necessity. But
other species would--at best--suffer a drastic drop in
numbers, and a few would vanish. The exact mechanism of
this disaster, though, depends upon the special role of the
estuary for a specific species. In general, the fauna of
an estuary can be sorted into four groups.

1. Truly Estuarine Species. Some species appear completely restricted to the estuary by a variety of factors,
not all thoroughly understood. The common oyster, for example, depends upon the estuarine plankton (microscopic,
free-floating plant and animal life) for food. But the
lower salinities also serve as a shield against enemies,
particularly the starfish and oyster drills, which cannot
tolerate salinities significantly below 20 ppt. And to
develop beyond the larval stage, oysters need clutch, the
solid material on the estuary bottom offered by the banks
of old shells from their ancestors or other shellfish.
Still other species, including the spotted sea-trout, lack
these highly apparent needs for an estuarine environment,
but they possess a special aptitude to survive in an
estuarine niche and are never found elsewhere.

2. Anadromous and Catadromous Species. Salmon, to
use a legendary example of an anadromous fish, spawn in
fresh water but mature at sea. Obstacles, such as dams or
severe pollution, in the estuary prevent the migration and,
typically, end the species' existence within that water system. Eels are the best known examples of catadromous species that live in fresh or brackish (slightly saline) water
during the adult stage but return to salt water for spawning. There are also variations upon these themes. Striped
bass, for example, spawn in fresh water, mature in estuaries, and spend an adult stage at sea.

3. Seasonally Estuarine Species. The striped anchovy,
spotted hake, and other species are estuarine residents on

a seasonal basis, consistently depending upon the estuarine environment to provide food and, in some cases, spawning areas during a significant part of the year. Many migratory shore birds--including the celebrated whooping crane--fall into this category. But this dependence appears to be lessening for a few species, such as the Canadian goose and mallard duck, as more flocks are wintering in farm lands.

4. Marine Species Using the Estuary as a Nursery. The most important species of marine resources in both value (shrimp) and weight (menhaden) for the United States are dependent upon the estuary to provide shelter and food during the juvenile stage of their development. Spawning occurs at sea, not in the estuary, but the adults would presumably not develop if the estuarine stage of the life cycle were thwarted.

Using the pink shrimp of Florida's west coast as an example, spawning occurs in offshore waters at depths of 100 to 150 feet, salinity between 36.1 and 17.7 ppt, and temperatures between 18° C. and 25° C. Within thirteen or fourteen hours, the eggs hatch and the larval shrimp rise to surface. At this point, they are about 100 miles from the coast, scarcely able to propel themselves and drifting almost helplessly at the mercy of winds and swarms of predators. Yet in some mysterious fashion they are able to gradually move towards the Florida estuaries, which they enter about three to five weeks after they hatched--but only an estimated 0.05% or less are believed to survive from eggs to this stage.

By the time the estuary is entered, the young shrimp have begun to swim, catch moving food, and have acquired tolerance to varying salinity and temperature conditions. After about nine months in the estuary, the shrimp have grown to commercial length of several inches and have returned to the sea where they complete their life cycle.

5. Food-Chain Dependence. Since estuaries nourish many of the primary consumers, such as menhaden and shrimp, of the total marine environment, they perform a vital function for the entire marine food chain. Tuna, porpoises, sea lions, seals, swordfish, and sharks, and similar secondary consumers feed directly upon menhaden and similar fish but also prey upon the menhaden's and shrimp's predators, such as the bluefish.

Impact of Change Upon the Estuary

Deep-rooted conflict in use of the estuaries is an issue that has not been fully comprehended by the United States society nor has it yet been approached with adequate resource management policies. Conflict arises from the intensity and variety of human use, often in conflicting and destructive fashion. Seven of the world's ten largest cities and fifteen of the United States' twenty largest are in the estuarine zone. Estuaries are transportation hubs where ocean vessels transfer their cargo to trucks and rail. Highways and rail lines radiate from the port cities along the rivers and estuarine coasts, often built on beds wrested from the estuary marshes. The estuaries' shores are lined with industries sustained by the surrounding commercial activity, transportation opportunities, ample water, inexpensive waste disposal, and the labor and markets of the estuarine city.

At the same time, estuaries are expected to nourish commercial fishing activity on the order of one-third of a billion dollars annually, and the estuarine environment is viewed as a recreational opportunity for swimming, boating, and sport fishing by the local urban citizenry. And the estuary is accepted as the urban dweller's window to wildlife--the birds, porpoises, seals, sea lions, and whales. Increasingly, too, the U.S. people are regarding the estuary and ocean as a component of their life-support system that helps to maintain the adequacy of their air and water quality in ways that are not fully understood. Estuarine factors that are causing the concern can be divided into two main geographic categories, those changes occurring in the watershed area upstream from the estuary and those within the estuary itself.

Changes in the Watershed

The estuary is the eventual recipient of virtually all wastes and side-products of man's terrestrial activities affecting water quality. Man can modify flow by diversions, spillways, dams, and levees. He can alter the water chemistry by nutrient enrichment and pollution with industrial chemicals, or he can upset the ecosystem's balance by introduction of new species. And he can add heat, the subject of the next chapter.

Modifications of Flow. Dams blocking the migration of anadromous fish, such as the shad, have long been the bane of environmentalists. But even when fish ladders and other devices for circumventing the dams have been provided, the spawning fish have often found that the deep, still, oxygen-deficient, and cold waters of the reservoir are not suitable for their eggs nor are the headwaters above the reservoir sufficiently passable. Dams prevent the movement of ice that once scoured some rivers, and dams also retain the sediment that previously passed into the estuary.

To operate hydroelectric dams efficiently, water should be released in equal quantities throughout the year. In areas where the major flow is in the spring, this means that the spring floods are retained until the reservoir is filled and then released evenly through the dry summer. In the estuary, spring flows are thus abnormally low but summer and fall flows may be increased. Hence, the estuary will be unseasonably saline during the spring spawning season, and the decreased currents will prevent the marked stratification and turbulent mixing required by some estuarine biota. For example, sardine production in the Nile estuary of Egypt has fallen precipitously with the construction of the Aswan High Dam.

Flood control dams are operated under a different philosophy. The objective is to retain just the flood peak that could prove dangerous, then to release the stored water as quickly as possible so the next flood threat can be met. But even this action prevents the periodic fresh-water enrichment of some estuarine marshes and creation of currents that would otherwise exist. Other changes have been caused by levees, such as those along the lower Mississippi, that send water down rivers with higher velocity, greater silt loads, altered currents, and scarce opportunities to create bordering marshes. Still other fresh-water flow changes are caused by ship channels and diversion canals.

On the ocean side of the estuary, modifications in the shoreline can affect the flow of salt water into the estuary. For example, ship channels through a barrier sand bar cut the littoral (parallel to shoreline) flow of sand, causing erosion downbeach unless dredges are continually passing the sand across the channel. Eventually, the sand and shore formations of the estuary are altered.

Except for anadromous species, however, the flow modifications do not typically devastate an estuary unless the fresh-water supply is cut off. These changes are, after all, similar to the annual variations that the estuary normally experiences. Adjustment in populations occur, some species declining in number and, possibly, others increasing. Since the new flow conditions are generally stable, recurring year after year, the ecosystem can reach a new equilibrium with most of the ecosystem's variety intact, although the new structure may not be as desirable as the original one.

Chemical Modifications. By contrast, chemical modifications can alter an estuary's ecosystem beyond recognition, and the changes are invariably undesirable by usual standards. The chemical change can vary from slight to drastic, from a low increase in a common ion--such as chloride leached from soil during irrigation--to a high concentration of an "unnatural" toxic chemical--such as a pesticide or petro-chemical by-product--from industrial processes. During their evolutionary development, the biota of the estuaries have not developed defenses to these assaults on their physiological systems.

The impact of chemical modification is rarely complete or spectacular at any one time since the vulnerability for some species will be higher than for others. For example, with low levels of chlorinated hydrocarbons, such as DDT, some species of clams appear to thrive but shrimp and certain other common fauna--and their predators--will disappear.

Within any one species, the vulnerability will also vary according to the life stage. DDT concentrations of 2 ppm (parts per million) can kill adult brown shrimp, but the juvenile shrimp will have disappeared long before these concentrations are reached. And chemicals do not always kill outright. Concentrations of 0.01 ppm of DDT irritate adult oysters and inhibit their shell growth, but some larvae will survive concentrations approaching 1 ppm. By biological magnification of the toxic chemical, low concentrations can also transform common primary consumers into living bombs for organisms higher on the food chain. Oysters exposed to 0.0001 ppm of DDT have accumulated flesh concentrations of 7 ppm, a magnification of 7,000 times. Yet generalizations are difficult to make since, for example, oyster drills have been fed diets of live oysters with 50 ppm of DDT without obvious ill effects or body concentrations.

Chemical changes can also produce subtle but important
ecosystem shifts that have unexpected results. A classic
case involved Great South Bay on Long Island, New York, in
1950. A combination of closing salt-water inlets into the
bay and the expansion of the duck farms on the bay's watershed led to a rapid increase in algae density and to a complete change in the type of algae present, the normal mixture of about three types being almost completely replaced
by a rarer species. The famous "blue-point" oysters, which
had flourished in the bay under the cultivation of descend-
ants from original Dutch families, gradually became unfit
for market and then died. Dissection showed that the oysters were unable to digest the new algae, since nitrogen
was being concentrated in a form inedible for oysters. In
effect, they were starving. Eventually, it was found that
the algae balance could be partially corrected by increasing
the water salinity with enlarged sea inlets. But this intensified starfish infestation of the oyster beds.

Changes within the Estuary

In the late 1960s, the U.S. Fish and Wildlife Service
estimated that half of the nation's estuaries had been
moderately modified, 23% had been severely modified, and
27% had been slightly modified. Alaska's estuaries improved
the statistics since none were considered to be severely
affected, while 62% of the Southwest Pacific's estuaries
fell into the severe category. (See Table 6-3.)

An inventory of estuarine changes along the North
Atlantic coast from Maine to Delaware during a five-year
period in the early sixties indicated that tidal marshlands
were being filled at an average rate of 4,500 acres annually
for:

Disposal of dredge spoil from harbor and channel improvement:	34%
Filled areas for housing developments:	27
Recreation developments:	15
Bridges, roads, parking lots, and airports:	10
Industrial sites:	7
Solid waste disposal:	6
Other purposes:	1
	100%

Table 6-3. Modification of the United
States' Estuaries

Biogeographic Zone	Degree of Modification of Estuaries (%)		
	Slight	Moderate	Severe
United States	27	50	23
North Atlantic	44	48	8
Middle Atlantic	5	68	27
Chesapeake Bay	44	50	6
South Atlantic	36	60	4
Biscayne and Florida Bay	50	50	0
Gulf of Mexico	15	51	34
Southwest Pacific	19	19	62
Northwest Pacific	13	50	37
Alaska	80	20	0
Hawaii	54	15	31
Great Lakes	35	46	19

Source: Fish and Wildlife Service.

The remarkable aspect of this survey is that the majority of the uses do not have any particular justification for occurring on estuarine lands. The use involving the largest percentage of loss--a spoil dump that buries estuarine marshes under the muds from the bottom of shipping channels--carried no significant social benefit. In most cases, the spoil could have been used to build islands almost as easily as laying the continuous blanket of mud. Also, the housing could have been placed on land at considerably less overall social cost, although the cost may have been higher to the individual developers.

Several frequent activities in the estuaries are highly disruptive, although they stop short of actually filling the marshes. Mining, especially for sand and gravel in the Northeast, has torn apart thousands of acres of estuarine marshland in the same manner that strip mining has devastated parts of Appalachia. Oyster-shell dredging in the South and Southwest, especially in Texas where permits were long regarded as political payoffs, has posed a similar political and ecological problem.

Mine tailings, the mineral material left after concentrated ore has been extracted, is sometimes washed or dumped into the estuarine zone where it smothers bottom organisms and clouds the water. One publicized case has been Lake Superior where a company has been discharging exceptionally fine tailings from iron-ore processing into the lake.[*] Because of Lake Superior's exceptionally vigorous currents, the fine particles appear to be scattering throughout the lake, which has until now miraculously escaped serious pollution.

The results of these and other man-made changes are difficult to determine since we have only a few imperfect statistics, such as commercial fish catches, and these are unrevealing because of changing technology and extension of fishing grounds. Some species, such as the croaker, have disappeared from the more polluted northern parts of their former habitat; and Connecticut has lost its once-thriving seed oyster industry. Yet these facts do not measure the overall productivity of the estuaries nor their importance for the ocean's biota. Nor do they begin to express the

---

[*] The Great Lakes are typically considered estuarine zones because they resemble the ocean's estuaries in most characteristics other than salinity.

aesthetic, scientific, and material loss that estuarine
degradation represents.

Estuarine Management Strategies

Political Background

In a conflict situation, "management" detached from
political confrontation is impossible. So for decades
estuaries have been subjects of intense political debate
and compromise, usually in favor of amply-financed commercial interests. The question of jurisdiction between local
(town and county), state, and federal governments has been
a continuing issue. With real estate taxes and industry at
stake, local governments have vigorously insisted that their
land-control powers be sustained. Traditionally, states
have claimed jurisdiction over all lands below the high-tide
mark, and their interest in the tidelands has been reinforced by mineral resources, especially oil but also sand
and gravel or, in the South, oyster shell. Yet the federal
government has ultimate responsibility for the nation's commercial and--to a lesser degree--sport fisheries (Department
of the Interior) as well as navigation (Corps of Engineers)
and interstate commerce (Department of Commerce).

Until the mid-sixties, the stalemated situation had
allowed local governments to hold the key powers except in
a few states, such as Connecticut and Maryland, where a
relatively weak state agency had limited jurisdiction over
the state-owned lands below high tide. The balance began
to shift in 1962 when Massachusetts passed the Massachusetts
Coastal Wetland Laws (Chapter 130, Section 27) that required
state licensing by the Department of Natural Resources for
all coastal wetlands projects. This power was expanded in
1965 to give the state the right to buy wetlands under emi-
nent domain proceedings when the state's powers were challenged. Other states, including New Jersey, Delaware, Connecticut, and Oregon, have followed in tightening state
control.

But the problem has remained acute in remaining states.
With pressure building upon the federal government to exercise control, new legislation began moving through the federal legislative system in the mid-sixties. Some powers
already existed under the Corps of Engineer's traditional
right to license all estuarine dredging and filling, but a

permit was usually granted if the project did not interfere with navigation even though a 1958 law required the Corps to obtain the "advice" of Interior's Fish and Wildlife Service. Each reform effort succumbed under the intense lobbying efforts of such widely assorted interests as state and local governments, land developers, dredge operators, and local land owners resenting the intrusion of the federal government upon their assumed prerogatives. After several years of debate, an Estuary Protection Bill emerged in 1968, but all real powers of land-use control, including easement, purchase, or permit systems, had been stripped away. Subsequent pleas by the Department of the Interior produced the Federal Coastal Zone Management Act of 1972 that provided planning grants to states and required federal actions in coastal areas to be consistent with approved state programs. Planning programs had been established in all but one of the thirty-four coastal states and territories by 1975.

General agreement exists among political observers that the basic problem is the lack of widespread public support for an estuarine management system. Many of the estuaries' features that need protection, such as the productive salt marshes, do not immediately and obviously affect public well being. To many, the estuary is remote "unused" land to be wrested from the greedy forces of the sea. Ecological studies are needed to identify in more detail the values to be protected, but even more important is the awareness that estuarine values are sufficiently important to save.

Alternatives

Estuarine management has two general objectives: (1) assure balanced consideration of all social needs when developers want to use the nation's limited estuarine resources and (2) restore the quality of those estuaries where the existing modifications cannot be adequately justified. The normal land-use policy tools described in Chapter 3 apply for a government agency seeking those objectives within the estuary. In addition, we have begun to develop technologies, such as artificial nurseries, for maintaining populations of some threatened estuarine species. We would be dangerously presumptive, however, if we believed that technologies could replace all the important functions of an estuary.

Restoring Quality

Improving water quality in the nation's streams is an important step in restoring the quality of estuaries. Eventually, however, restoration will also require some corrective measures within the estuary. Sometimes this will simply mean breaking apart the spoil levees that prevent adequate salt-water access to the marshes. In other cases, removal of old bulkheads (walls) and replanting of estuarine flora will be necessary. In some harbors, the recent shift to containers for marine cargo has made many old docks obsolete and may have facilitated the restoration of earlier estuarine conditions.

Estuarine Resiliency

The resiliency of the estuary should not be underestimated. Estuarine flora and fauna are adjusted to abrupt and frequent changes in their physical environment. In nature, salinities change, temperatures rise and fall, and currents shift--frequently without drastic biological losses. Some organisms have even become resistant to pesticides, and others may acquire this resistance in the future. But the drastic and shifting character of man-made change in the estuary reduces the effectiveness of this natural resiliency. No fish can grow on filled-in marshes. We cannot expect to resurrect the pesticide-resistant fish that died when their food, the shrimp, were poisoned by the pesticides. But these modifications will continue to occur until the public has developed an awareness of the estuary's importance and appropriately strong legislation has been enacted and implemented.

QUESTIONS FOR DISCUSSION

1. List the physical and biological characteristics--especially environmental tolerances--that you believe estuarine fish generally need for survival. Fresh-water fish. Salt-water fish.

2. Why is the bottom strata of an estuary important for biota? Give examples.

3. Why do you think vast expanses of the oceans are void of significant life?

4. Explain why turbidity--or cloudiness--of an estuary or lake changes its biological characteristics.

5. Describe a "public awareness campaign" that you would consider desirable before introduction of new estuary protection legislation in a state legislature.

6. How would you explain the value of estuaries to (a) a real-estate developer owning water-front property, (b) a local town-council representative concerned with rising tax rates, and (c) a state highway official considering the estuary's shore for a new expressway?

SUGGESTED READINGS

Between 1964 and 1972, four particularly informative symposia on estuaries were held, and the papers were subsequently published. The most important of these proceedings--and the most complete reference book on estuaries now available--is being distributed by the American Association for the Advancement of Science.

1. G. H. Lauff (ed.), Estuaries, Publication No. 83 (Washington, D.C.: AAAS, 1967).

2. R. F. Smith, A. H. Swartz, and W. H. Massmann (eds.), A Symposium on Estuarine Fisheries, Special Publication No. 3 (Washington, D.C.: American Fisheries Society, 1966).

3. Marine Technology Society, Tools for Coastal Zone Management, Proceedings of the Conference, February 14-15, 1972 (Washington, D.C.). Available at the Marine Technology Society, 1730 M Street, N.W., Washington, D.C., for $12.00.

4. Bostwick H. Ketchum (ed.), The Water's Edge: Critical Problems of the Coastal Zone, Report of the Coastal Zone Workshop of May 22-June 3, 1972 (Cambridge, Mass.: MIT Press, 1972).

In 1970, the U.S. Department of the Interior issued their seven-volume report on the estuaries. Their approach is managerial, and the material tends to be wordy and imprecise, although specific estuaries are discussed. In the same year, the National Academy of Sciences and National Academy of Engineering also issued a report.

5. U.S. Department of the Interior, National Estuary Study, Volumes 1-7 (Washington, D.C.: U.S. Government Printing Office, 1970).

6. NAS-NAE, Wastes Management Concepts for the Coastal Zone: Requirements for Research and Investigation (Washington, D.C.: NAS-NAE, 1970).

A modest, slightly outdated but still helpful and readable paperback is:

7. John Clark, Fish and Man: Conflict in the Atlantic Estuaries, Special Publication No. 5 (Highland, N.J.: American Littoral Society, 1967). Obtainable for $1 from the American Littoral Society, Highland, N.J. 07732.

A few textbooks specifically directed towards coastal zone management are beginning to appear. One example is:

8. James C. Hite and J. M. Stepp (eds.), Coastal Zone Resource Management (New York: Praeger Publishers, 1971).

ESSAY

Oil Pollution[*]

The Ubiquitous Pollutant

Many mysteries shroud oil pollution. We have only the crudest guess of how much occurs, and just a small fraction is ever seen by land-bound humans. While oil is the most apparent pollutant of the oceans, which cover most of the world's surface, oil's mechanical and chemical properties vary widely according to its source, degree of processing (if any), temperature, and sea conditions. We know petroleum is toxic, but generalizations are hazardous because it consists of many compounds, which, in turn, form other compounds in the environment. Only now are we beginning to realize vaguely how toxic some of the compounds are, especially in their long-term effects. Yet some ocean organisms seem stubbornly unaffected.

Sources

The quantity of oil spilled into the marine environment is probably on the order of a billion gallons per year, about one-thousandth of the amount being extracted from the earth's substrata. Oil pollution seems to accompany all technological activities, creating countless small sources of spillage. Most pollution of the seas, however, is associated with transportation of oil from the wells to industrial centers. Ships are held responsible for the major proportion because of spillage during (a) adjustment of the ship's weight or ballast, (b) loading and unloading, and (c) accidents.

Until recently, much of the pollution was caused by a routine operation. Crude-oil tankers cannot sail empty since the ship would be too light, float high in the water, and rock unstably in the water. Propeller and rudder would be partially exposed to the air and be dangerously ineffective. In the past, the solution was to pump sea water into

---

[*]For a knowledgeable discussion of oil pollution and its control, see: A. Nelson Smith, Oil Pollution and Marine Ecology (New York: Plenum Press, 1973).

133

the oil tanks for weight until the ship arrived at a port where more oil could be loaded. Since some oil always clings to the inside of the tanks, the tanks had to be cleaned first so oily water would not pollute the harbor when oil was loaded. Thus, about four-hundredths of every oil shipment was pumped overboard in the tank washwater.

Because of harsh criticisms and sanctions by some countries, ship operators have begun using a special tank to store the washwater and separate most of the oil for recovery. This is known as LOT for "load-on-top," referring to the loading of fresh crude oil on top of the water-oil dregs in the tanks. Over three-quarters of the oil tankers now have these storage tanks, but this does not seem to have drastically reduced the amount of oil on the oceans.

Marine accidents account for the largest single instances of spillages. About half of all sea traffic consists of tankers, and the law of averages appears to dictate that between fifty and one hundred serious incidents involving tanker damage and some oil spillage occur each year. When the "Torrey Canyon" struck rocks near the English Channel in March, 1967, some thirty million gallons of oil spilled into the sea. During the following year, the "World Glory" suffered hull failure and poured another fifteen million gallons into waters off South Africa. Economies in designing, constructing, and operating the ships enhance the possibility of accidents. Only a single sheet of hull steel separates oil from the ocean waters. Single propellers are the most economic, but they reduce the maneuverability and stopping capacity of tankers. Crew training has been criticized, and communication within the ships is regarded as poor.

Accidents at offshore oil wells are considered the next largest contributor to petroleum pollution of the marine environment. Over 10,000 wells have been drilled in the Gulf of Mexico, mostly off Louisiana; and another 1,000 have been drilled along the California coast. With the emphasis on energy self-sufficiency, the Atlantic coast of the United States is expected to be the next scene of extensive drilling. "Blow-outs," the loss of control over the oil flow at high pressures, represent one of the dangers. Between 20,000 and 200,000 gallons per day of oil were lost in a blow-out at Santa Barbara in January, 1969, and an estimated 3,250,000 gallons were believed to have polluted the coastal area during the first hundred days. An explosion at a well in Louisiana during April, 1967, caused about 40,000 gallons per day to flow into the sea.

In addition, the problems of moving the oil to shore facilities or tankers present constant opportunities for spills, either through failure of equipment or handling errors. Just loading oil for fueling cargo ships poses risks, and spillage by small fishing and sports boats also provides a share of the oil pollution, especially in densely populated estuarine zones.

On shore, most refineries use considerable water for cooling, and oil leakages into the cooling water frequently occur. Despite apprehensions over the possible effects of Alaskan pipeline construction, pipelines have not proven serious polluters, although spills do occasionally occur. One estimate is that one to two spills each year for every 1,000 miles of pipeline should be expected. Numerous other oil-pollution sources could be cited. Used crankcase oil has been increasingly dumped into sewers leading to rivers and estuaries. Spillages and effluents from chemical industries provide a share, and failures of storage tanks have caused some devastating spills. All combustion of oil is usually incomplete, leaving some petroleum residues to settle on land or water. But these sources are relatively minor.

Effects

Spills can involve either crude oil or the refined products, but most of the major spills have involved crude. Crude oil from different sources has generally the same ingredients, but proportions differ. Some can be thick and others, thin. All, though, have toxic ingredients. Among the more troublesome are the lighter compounds, especially the polycyclic aromatic compounds (PAH) that include many known carcinogens and can combine with common environmental materials to form other carcinogens.

In a typical oil spill, the mechanical effects of thick ooze are the most obvious. Sea life on the surface is coated and is either asphyxiated or immobilized. Toxic effects are more subtle. Fish eggs and larvae, which typically float near the surface, are generally killed quickly. In confined space, other sea life will often die though over longer periods. The season will affect the toxicity, and species will differ widely in susceptibility. Sometimes

135

the mechanisms are multiple. For example, diving sea birds that must plunge through the oil and rise again from beneath it will quickly become coated. When they preen to remove the oil, they will ingest large quantities; and the oil in their feathers will affect their ability to swim, fly, and maintain body warmth. In other words, they will succumb to the combined effect of poisoning, freezing, and drowning.

Solutions

Ironically, economics--the same force that stimulated development of the problem--may be furnishing part of the solution. Spilled petroleum represents financial loss, and additional alertness to preventing spills can be expected as prices rise. The ultimate solution is to replace the use of oil by other energy sources. Meanwhile, international controls are slowly pressuring tanker operators to decrease practices that spill oil. Britain passed their first Oil in Navigable Waters Act in 1922, and the U.S. followed with the Oil Pollution Act of 1924. Both countries have been joined by others, and the legislation has been constantly strengthened. Yet control over individual actions, especially carelessness in isolated seas, is difficult to achieve.

Once oil has been spilled, numerous devices for skimming or soaking up the oil from the water's surface have been devised. None really succeeds in stormy weather or other adverse conditions. In coastal areas, straw is still being used to absorb oil. When conditions are appropriate--which is rarely--materials can be added to oil to cause sinking. In other cases, solvents can be added to dissolve the oil, though solvents tend to be more toxic to sea life than the oil itself.

In Perspective

From one perspective, concern for oil pollution is easily shrugged off. Related problems--scarcity of energy alternatives, control of ocean resources, and the festering imbalance of payments with oil-producing countries--seem overpowering in importance. Furthermore, use of oil is inescapable in our present technological society. Yet, from another perspective, oil pollution is the least excusable because none of the major sources is necessary in oil use but is related to human callousness, carelessness, or ignorance.

Chapter 7

THERMAL POLLUTION

> "Heat not a furnace for your
> foes so hot that it do singe
> yourself." William Shakespeare
> (1613)

Problem Elements

Waste heat may seem an insignificant example of water
pollution. After all, man's annual release of stored heat
is a minute fraction--less than one thousandth--of the sun's
energy striking the earth's surface each year. Yet by the
sheer volumes of discarded heat that are concentrated at
single points, we have been able to produce many of the
common characteristics of pollution--fish kills, obnoxious
odors, cloudy waters, disagreeable tastes, and other disruptions of natural ecosystems.

Public Health Service statistics indicate why the electric power industry has been singled out as the main culprit.

Table 7-1. Use of Cooling Water by U.S. Industry, 1964

Industry	Cooling Water Intake (Billions of Gallons)	Percent of Total
Electric power	40,680	81.3
Primary metals	3,387	6.8
Chemical and allied products	3,120	6.2
Petroleum and coal products	1,212	2.4
Paper and allied products	607	1.2
Food and kindred products	392	0.8
Machinery	164	0.3
Rubber and plastics	128	0.3
Transportation equipment	102	0.2
All other	273	0.5
Total	50,065	100.0

Source: U.S. Department of Commerce, 1963 Census of Manufacturers.

Excluding irrigation, over half--52%--of the estimated water withdrawals in the United States during 1970 were for absorption of waste heat from steam electric utilities. Another 10% was for cooling industrial processes.

Three trends threaten to compound the problem.

(1) Despite rising prices, increases in electricity consumption are expected to continue for the immediate future, although perhaps not at the past rate of about 7% annually, a rate that has meant doubling of electricity use each decade for over thirty years.

(2) Nuclear generated power, which is now more wasteful of heat than fossil-fuel generation, is expected to expand its 4% share of generating capacity in 1970 to a larger proportion, although probably not to the about 40% share in 1990 earlier estimated by the Federal Power Commission.

(3) Thermal generating plants are becoming larger with commensurate discharges of waste heat. The average plant constructed in the 1950s had less than 300 megawatts capacity, but the average plant contracted for construction in 1968 was rated at 835 megawatts and several exceeded 1,000 megawatts.

These trends towards larger, often nuclear plants can be challenged. Unless older, less desirable plants are being replaced, the need for the additional quantities of electrical energy can be questioned in terms of the contribution to quality of life. Economies of scale are possible in large plants and control of side effects, such as air pollution, can be maintained more easily; but the sheer size of the operations raise concerns involving plant safety, transmission losses and hazards, concentration of pollutants, and--in the case of nuclear plants--security of radioactive materials. Nevertheless, power plants and industries requiring water for cooling are being constructed, and maintenance of water quality means that the physical, chemical, and biological characteristics must be protected from changes caused by thermal pollution. Before considering means of controlling thermal pollution, the effects of heat on water characteristics should be considered.

Table 7-2. Water Properties

Temperature ($^\circ$C.)	($^\circ$F.)	Density (gm/cm^3)	Viscosity (centi- poises)	Vapor Pressure (mm Hg)	Dissolved Oxygen Saturation (mg/l)
0	32	0.99987	1.7921	4.58	14.6
0	39.2	1.00000			
5	41	0.99999	1.5188	6.54	12.8
10	50	0.99973	1.3077	9.21	11.3
15	59	0.99913	1.1404	12.8	10.2
20	68	0.99823	1.0050	17.5	9.2
25	77	0.99707	0.8937	23.8	8.4
30	86	0.99567	0.8007	31.8	7.6
35	95	0.99406	0.7225	42.2	7.1
40	104	0.99224	0.6560	55.3	6.6

Source: U.S. Department of the Interior, FWPCA, Industrial Waste Guide on Thermal Pollution.

Physical Changes

If you took some water and poured it, weighed it, determined the oxygen in it, and made other measurements before and after heating it, you would find that the addition of heat makes some distinct changes though detection of these changes may require delicate instruments.

Density. Density, which is measured by weighing a defined volume of water, changes with temperature. A gram was historically defined as one cubic centimeter of pure water at 4° C. (39.2° F.). As heat is added to water, it becomes less dense, weighing less per unit volume. Hence, warm water tends to rise and form a layer on top of cold water.

Viscosity. Just as with syrup, addition of heat to water makes it less viscous, more freely flowing. In nature, this results in the tendency of warmer waters to flow more quickly than colder, the warmer often forming a channel over the colder layers.

Vapor Pressure. We also know that heated water tends to evaporate, and this is measured by the increased vapor pressure over water when heat is added. So we can expect to lose more water by evaporation from lakes and rivers that

are heated, a significant factor in some arid parts of the
country.

   Dissolved Oxygen. A more critical factor for fish is
the lower dissolved oxygen in heated water. When saturated,
water at 35° F. will hold roughly twice as much free oxygen
as the same water heated to 100° F. And quantities of other
gases normally present in trace amounts, such as nitrogen,
may become unstable in water that is rapidly heated. This
phenomenon can affect aquatic life for miles along a river
that is being heated by a generating plant. (See Figure
7-1.)

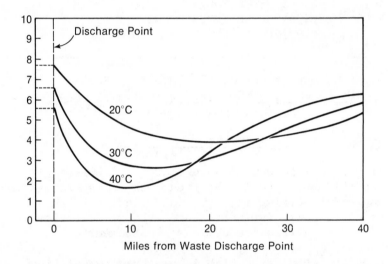

Figure 7-1. Typical effects of water temperature
            upon dissolved oxygen content.

Interaction with Other Mechanisms

   In a natural water system, many of these effects may
be camouflaged or overridden by combined or grosser effects.
Mixing is always occurring. Water in motion is diverted by
any physical boundaries, such as rocks. Chemical oxidation,
photosynthesis, diffusion, and mixing is usually more important than theoretical saturation levels for oxygen in
determining the amounts of free oxygen available for fish.
Yet heat effects can become critical, especially for the
more sensitive aquatic life or vulnerable water bodies.

The potential impact of adding heat to a deep, slowly
flushed lake has been extensively described with the example
of Cayuga Lake, one of the Finger Lakes of Central New York
State and the site of Cornell University.[*] During the winter, the lake's waters become almost uniform at 35-40° F.,
and mixing occurs. Dissolved oxygen and nutrients are distributed to all layers. In the spring, the lake begins to
stratify into two distinct layers, the epilimnion (the upper
strata) and the hypolimnion (lower strata) that are divided
by the thermocline, a thin layer of abrupt temperature
change that virtually blocks mixing.

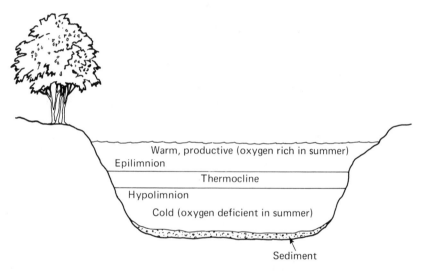

Figure 7-2.   Thermal profile of a deep lake in late summer.
(From Mitchell, Introduction to Environmental
Microbiology, Prentice-Hall, Inc., 1974.)

By late summer, the epilimnion extends thirty-five to
fifty feet below the surface and is intensely active biologically since it receives energy from sunlight and oxygen
from both photosynthesis and the air. Meanwhile, however,
the hypolimnion is losing its reservoir of oxygen through
respiration of living organisms and the decomposition of
dead organic materials that sink from the epilimnion. And
nutrients are building up in the hypolimnion because of the
biological breakdown of dead organisms.

---

[*]See Dorothy Nelkin, Nuclear Power and Its Critics: The
Cayuga Lake Controversy (Ithaca: Cornell University Press,
1971).

If this trend were not reversed, the hypolimnion would become "dead" because of oxygen deficiency while the epilimnion would be starved for nutrients--except by natural enrichment and man's wastes flowing from the outside. Fortunately, the epilimnion begins to cool in October and gradually shrinks in depth. By November the thermocline has reached the lake's surface and been broken by winds. The lake's waters again mix with the hypolimnion being recharged with oxygen while the upper layers receive their annual nutrient supply.

At Cayuga Lake, scientists became upset with a proposal to construct a large nuclear power generating plant on the lake's shores. Water was to be drawn from the hypolimnion, where temperatures are always below 45° F., and released on the epilimnion. For the power company, the cold hypolimnion waters meant exceptional efficiency in the plant's operation, but scientists became concerned that the discharge of heated waters at the surface would prolong the thermal stratification of the lake and, thus, cause oxygen exhaustion in the hypolimnion. Furthermore, nutrients drawn from the hypolimnion would be discharged into the epilimnion during the summer, aggravating the lake's already serious eutrophication. Between 1928 and 1965, transparency in the lake had declined from about seventeen feet to eight feet.

Chemical Effects

Both chemical and biochemical reactions quicken with rises in temperature. In general, the speed of a chemical reaction approximately doubles for each 10° C. (18° F.) increase in temperature. Thus, dissolved oxygen can be depleted more rapidly by simple chemical reactions that oxidize common industrial materials, such as iron particles.

Microbial activity is also hastened. Most chemical reactions caused by micro-organisms occur through catalytic actions involving enzymes, which are temperature sensitive. As noted in the discussion on sludge digestion, many--if not most--organisms affecting water quality reach maximum activity at 86° to 99° F. Therefore, the oxygen sag curves for streams may be shorter and more abrupt with higher temperatures than lower.

142

Biological Effects

We know from examination of ecological systems that
various organisms specialize in different sets of physical
conditions, temperature being one of the more important.
If we vary temperature, an organism will eventually find
itself less efficient at survival than other organisms and,
finally, a directly lethal temperature is reached. For
example, brook trout will die at 73-77° F. (23-25° C.), but
they become slow at catching minnows at 63° F. (17° C.) and
incapable of catching any at 70° F. (21° C.) so they can
starve before actually succumbing to heat. And, because
the body processes are accelerated by heat, the trout can
starve sooner in warm water than cool.

Most aquatic animal life is classed as poikilothermic,
or cold-blooded, because their bodies adjust to the environment's temperature. In most fish, the body temperature is
only 1° to 2° F. (0.5° to 1.0° C.) higher than the surrounding water. With this built-in inability to resist temperature changes, fish life is profoundly affected by temperature changes in at least four ways. It should be noted,
though, that some aquatic life can adjust to variations.
For example, the aurelia, a species of jelly fish caught
near Halifax, Nova Scotia, has an upper lethal temperature
of 27-28° C., but the same species from Florida has an upper
limit of 40° C.

(1) Metabolism. Metabolic activity can double or even
triple with an 18° F. (10° C.) increase in temperature.
Respiration increases to a point usually a few degrees below
lethal temperature. Death may be caused by (a) changes in
enzyme activity, (b) melting of cell fats, (c) coagulation
of cell proteins, (d) reduction and permeability of cell
membranes, and (e) toxic materials produced by metabolism.

(2) Reproduction. Many aquatic organisms can survive
over a broader range of temperatures than they can accept
for successful reproduction. The oyster, for example, has
been found thriving in the effluents of power plants at
temperatures far above those acceptable for spawning. Incu-
bation of fish eggs and development of young fry are usually
similarly sensitive to temperature. Nor are the temperature
bands always equivalent. Salmon eggs, for instance, have
been hatched at temperatures that proved lethal for the fry.
The impact of heat can be enhanced by different levels of
salinity in estuarine areas.

Table 7-3. Maximum (Provisional) Temperature
Considered Compatible for Various Fish Species

93 F.: Growth of catfish, gar, white or yellow bass,
spotted bass, buffalo, carpsucker, threadfin shad,
and gizzard shad.

90 F.: Growth of largemouth bass, drum, bluegill, and
crappie.

84 F.: Growth of pike, perch, walleye, smallmouth bass,
and sauger.

80 F.: Spawning and egg development of catfish, buffalo,
threadfin shad, and gizzard shad.

75 F.: Spawning and egg development of largemouth bass,
white, yellow, and spotted bass.

68 F.: Growth or migration routes of salmonids and for
egg development of perch and smallmouth bass.

55 F.: Spawning and egg development of salmon and trout
(other than lake trout).

48 F.: Spawning and egg development of lake trout, walleye,
northern pike, sauger, and Atlantic salmon.

Source: U.S. Department of the Interior, FWPCA, Report of
the Committee on Water Quality Criteria.

(3) Development. Each fish seems to have optimum development temperatures for maximum growth, but these may
not be the same as optimum temperatures for rapid growth.
For example, minnows (Cyprinodon maculatus) have been found
to grow more rapidly at 30° to 35° C. than at 15° to 20° C.
However, after a year, fish raised at 20° C. lived longer
and surpassed the length of those from 30° to 35° waters.

(4) Synergistic Effects. Scientists are frequently
confounded when assumptions based on laboratory tests with
aquatic species are transferred to the natural environment
where more factors may be existing. Heat and toxic substances are frequently more lethal than either factor alone.
For example, a 15° F. (8° C.) rise in temperature triples

the toxic effect of O-xylene. Furthermore, fish are more plagued with disease and parasites at higher temperatures. A fungus disease, _Dermocystidium marinum_, in oysters has been related to the severity of winter temperatures and the warmth and duration of summer temperatures. When water temperature reaches 25° C., the fungus spreads rapidly and entire beds of oysters can be wiped out in four to six weeks.

Adaptability to Change

Fish species, especially younger fish, are more adaptable to slow and gradual changes than sudden ones. The most spectacular temperature-caused fish kills in the United States have been during sharp falls in temperature in bays on the Gulf of Mexico. Some of the more disastrous direct fish kills from electric generation plants and industrial cooling operations have been caused by thermal shocks resulting from one of two conditions.

1. Sudden discharge of heat. An electric generating plant suddenly starting, especially at the height of summer, can push temperatures quickly into the lethal range without giving fish an opportunity to escape.

2. Sudden closing of discharge. In winter, electric power plants' outfalls have been found to be favorite refuge areas for many species, which seek their preferred temperatures along the heat plume in the colder natural waters. A sudden end to discharges throws this artificial ecosystem into chaos, causing fish kills if the background water temperatures are sufficiently low.

Like the fish, each algae species or groups of species appear to have a particular temperature range that is optimum and a broader range that can be tolerated. For instance, diatoms in an unpolluted stream will thrive with little competition at 64° to 68° F. (18° to 20° C.), green algae at 86° to 95° F. (30° to 35° C.), and blue-green algae at 95° to 104° F. (35° to 40° C.). Unfortunately, the blue-green algae is considered the least desirable and is widely associated with eutrophication.

In any analysis, the water biota should be viewed as elements in a total system. A lone intrusion, such as the addition of heat, rarely remains an isolated phenomenon but, for example, eliminating an algae can distort the balance of all remaining organisms--algae, bacteria, insects, fish, birds, and other links in the food chain--although the changes may not be immediately discernible to a casual observer.

Control of Thermal Discharges

Limiting Electrical Demand

As noted before, the most effective means of solving a problem can be to prevent the problem from arising. If additional electricity is not generated, additional thermal pollution will simply not exist to be considered a problem. For numerous social reasons, we will not want to prevent all increases in electrical generation, but it is equally apparent that a constant growth rate in "demand" is not sacrosanct. The need for additional electrical generating capacity should be subjected to the same benefit-cost considerations that are beginning to be applied to other demands upon scarce resources.[*]

Once the use pattern for electricity in an area is known, the approach to limiting use must depend upon individual circumstances. Usually the rate structure is the key element. Rates are typically lowest for the large user and, particularly, the large, off-peak (normally late night and early morning) user. But this has encouraged industrial use that extends into peak hours, increasing the need for new generating facilities and penalizing the domestic user with costs, both monetary and environmental, and the threat of power interruptions. Most energy in densely populated areas is consumed by commerce and industry; and sometimes a single industrial plant, such as an aluminum refinery, can consume more than half of the output from a local generating facility.

In other cases, electrical demand is raised because of special tax, cost, or income advantages to a particular class of consumer. Modern office buildings, which tend to

---

[*] Besides producing thermal pollution, power generation--both nuclear and fossil fueled--causes losses of fuel reserves, scarring of natural landscapes through mining, usurpation of irreplaceable open space by generation and transmission facilities, and pollution of air and water by various pollutants other than heat.

be gigantic glass boxes with high heat losses in winter and
heat gains in summer, almost invariably have electrical
cooling and, sometimes, heating with the cost justified because of tax advantages, low initial costs, and minimum
maintenance expenses. Modifications in design and construc-
tion--smaller glass surface area and improved insulation--
could diminish electrical use.

Improving Existing Technologies

In general concept, a steam-generating plant is a relatively simple device. A generator, which resembles a huge
electric motor in appearance and construction, is turned by
a turbine, a propeller-like device run by the force of high-
pressure steam on one side of the blade and low pressure on
the other. The process begins when the water is heated in
a furnace by a fossil fuel or nuclear reaction, becomes
high-pressure steam, and is carried to turbines. After
spinning the turbine, the steam's temperature will have
dropped from over 1,000° F. to less than 100° F. Steam
pressure will have fallen from 2,000 or more pounds per
square inch to something less than atmospheric.

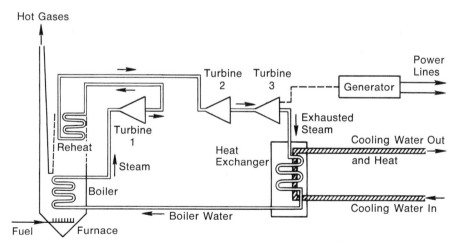

Figure 7-3. Basic elements of a typical
steam power plant.

This low-pressure, low-temperature steam then goes to
a heat exchanger (condenser), which is simply a series of

tubes submerged in cooling water, where the steam is converted back into water before repeating the cycle. When
steam converts to water, heat energy is released; and this
heat must be removed or incoming steam will remain steam.
The heat removed from the heat exchanger is the source of
thermal pollution. In fossil-fuel plants, about 15% of the
energy input is lost in the heating of the water, and this
waste heat is ejected directly into the atmosphere through
the smokestack.

In a nuclear-generating plant, the smokestack does not
exist; and the in-plant heat losses are usually assumed to
be just 5% of the energy in the fuel. Thus, a nuclear plant
emits a considerably larger proportion of its waste heat
into the cooling water than a fossil-fuel plant, a factor
contributing to the nuclear plant's 50% or more higher
thermal pollution potential despite a difference between
their efficiencies of only about 7%. For example, a fossil-
fuel plant operating at 40% efficiency will discharge 3,800
BTU for each kilowatt hour generated while a nuclear-fuel
plant attaining 33% efficiency will discharge 6,400 BTU into
local waters.

Historically, the efficiency of the fossil fuel-steam
generating technology has been spectacularly improved. In
1930, the average thermal efficiency of steam-electric
plants in the United States was about 17%, and an estimated
13,420 BTU had to be discharged into the water environment
for every kilowatt hour generated. By 1966, the thermal
efficiency had risen to nearly 33% and the discharge of
waste heat had declined to 5,440 BTU per kilowatt hour.
In the more efficient plants, the efficiency rate exceeded
40%. But further spectacular improvements in efficiency
cannot be expected.

Under the constraints of the Second Law of Thermodynam-
ics, a heat engine using a Carnot cycle operating ideally
with steam at $1,000^{\circ}$ F. and condenser cooling water at 50° F.
can attain a theoretical maximum efficiency of 65%. But
heat losses in the steam pipes, turbine, and other equipment
inevitably lower the efficiency. The theoretical limit can
be raised slightly above 65% by (a) increasing the steam
pressure in the boiler, (b) lowering the pressure behind
the turbine, (c) increasing the temperature of the steam,
or (d) refining the cycle by several techniques, such as re-
heating the steam after it leaves the turbine but before
entering the condenser. But all these devices have been
used in the modern generating plant, and further significant

improvements would involve temperatures and pressures above the safe operating limits of modern metallurgy.

Table 7-4. Steam-Electric Plant Statistics-- National Average Basis

Year	Heat Rate (BTU/KWH)	Thermal Efficiency %	Waste Heat to Cooling Water (BTU/KWH)
1930	19,800	17.24	13,420
1940	16,400	20.81	10,530
1950	14,030	24.33	8,510
1960	10,760	31.72	5,730
1962	10,558	32.33	5,560
1964	10,462	32.62	5,480
1966	10,415	32.77	5,440

Source: Federal Power Commission, Steam Electric Plant Construction Cost and Annual Production: 1966.

New Generating Technologies

Another approach would be to introduce a new generating technology that would discharge minimum waste heat directly into the atmosphere without the obnoxious gases found in combustion of fossil fuels. Many proposals are constantly being made, but most are merely gleams in a few eyes--and will remain such. Most of the concepts fall into three categories.

1. Many hopes have revolved about invention of a new generation technology that would be used with the existing, possibly nuclear, types of heat sources. A leading candidate is magneto-hydrodynamic (MHD) power generation, which substitutes hot gases moving rapidly through a magnetic field for the conventional rotor-stator generator. With single cycles, thermal efficiencies of 50-55% could be achieved, and this could be raised to 60-70% with a binary cycle. Despite considerable research both in the United States and the Soviet Union, however, a number of engineering problems are still considered to be unsolved and a reasonably reliable pilot model has not yet been constructed.

A related technology is electro-gas dynamics, the production of power by converting energy of flowing gases directly into electricity. No moving mechanical parts would be necessary in a working model. Much to the sorrow of its devoted advocates, however, the major research effort in this field under the sponsorship of the Office of Coal Research was terminated a few years ago.

2. Even simpler would be a method of generating electricity directly from an energy source without having to heat an intermediary material, such as a gas. Most hopes have centered upon the fuel cell, which produces electricity through chemical reactions assisted by a catalyst. Working fuel cells have been developed for the space program, but their cost has been exorbitant, and the energy output has been relatively small. If a technological breakthrough can be found, it is believed that a fuel-cell system would have exceptionally high efficiency, possibly 60 to 70%, would not require any cooling water, and would not produce air pollution. One research effort is focusing on use of natural or coal gas, which is already piped into homes for space and water heating. Electricity needs could be supplied by simply adding another home appliance.

In the far future, some scientists believe fusion power, the same process used by the sun and in hydrogen bombs, can be harnessed to produce electricity without any serious worry about future energy resources. But the engineering problems of producing and controlling an extended fusion reaction with all the attendant heat and radiation problems have proven insurmountable to date. Even if these difficulties are overcome, some skeptics believe that a fusion plant would be impossibly radioactive and dangerously unstable for widespread use.

3. If a natural source of energy similar to the falling water of hydroelectric developments could be exploited, much of the concern over efficiency and pollution would be eased. Three possibilities, geothermal, solar, and wind energy, have been discussed. Geothermal steam, naturally occurring deposits of subterranean steam, have already been harnessed for electricity in California and in a few other countries. Advocates believe that the geothermal beds underlying many of the western states could supply most of the electrical needs for local populations, an important point because cooling water is especially scarce in this region. In areas where geothermal beds do not naturally occur, such as in the Northeast, there have been proposals to experiment

150

with insertion of water into deep wells, and it is possible that a breakthrough may occur in this concept.

Solar energy, of course, is one of the more obvious and rudimentary of the natural energy sources. However, cloud cover, seasons, and nightfall have made conversion so unreliable and intermittent that it has only been used for space heating in a few instances and water distillation in particularly barren places. (Hydroelectric power, of course, is indirectly provided by solar energy that evaporates water so that it can later fall as precipitation and be collected behind dams.) Research has focused on two approaches: (1) capturing solar energy as heat and (2) direct conversion into electricity. Space heating of buildings by circulating heated water from a rooftop panel to rooms or utilizing the greenhouse effect directly has been used experimentally and wider application appears inevitable. Except in the space research program, direct conversion to electricity has not been considered feasible because of the high cost of manufacturing conversion cells. The situation may change, however, because of recent innovations in cell manufacturing.

Wind has been harnessed as an energy source in the past, and the solitary windmill on the horizon is still occasionally seen. A few examples of electrical generation by wind can be found, but the unreliability of wind and the huge size of windmills needed to replace conventional energy sources have completely discouraged development. This may change, though, as the concept of supplemental energy sources is accepted and the high cost of fossil fuels makes the capital costs of wind conversion more attractive. Conversion of tidal energy into electricity has also lagged because it is an intermittent source and, more critically from an environmental viewpoint, it can drastically affect estuarine ecosystems. Other potential sources, such as the difference in temperatures between water layers in oceans, have been discussed in technical literature but not explored vigorously.

Nuclear fuel's primary advantage has been its unusual compactness. One pound of uranium has the energy equivalent of about 1,500 tons of coal. Yet the existing commercial reactors use only about 0.5% of this energy for heating the boiler water, 95.5% being wasted, since only a rare uranium isotope can now be used. Because ore containing the needed uranium isotope is rapidly being depleted, a "breeder" reactor design is being sought so that a nuclear fuel can be reused. But this still will not reduce the water comsumption for carrying waste heat from condensers. It is possible, though, that new plant designs using exotic liquids, such as liquid sodium or helium, and higher pressures will enable nuclear plant efficiencies to rise and the coolant water demands to drop.

Heat Utilization

Superficially, thermal pollution presents a strange situation. The nation spends millions of dollars heating houses yet enough waste heat is discharged from generating plants to heat all the houses in the United States. Justifiably, heat utilization has become a popular subject among engineers and environment-minded dreamers. But the most appropriate uses of waste heat have not yet been determined, nor have the management institutions been established to handle utilization.

Utility companies are notoriously provincial in their thinking; and the franchise system, a government-licensed monopoly with guaranteed returns on investments, does not give the utility companies a high incentive to innovate. This lack of excitement within the industry has meant that the more imaginative, creative managerial types tend to reject employment with electric utilities, although the recent energy crisis may be changing the situation.

Current heat utilization proposals can be classified into six or more categories: (a) space heating and cooling, (b) aquatic farming, (c) agriculture, (d) water conditioning, (e) maintenance of shipping lanes, and (f) various industrial applications. Except for some forms of water conditioning, each use would require diversification of a power utility's attention and, probably, investments. Thus, the human problems--including the social and political-- appear the most formidable. Yet there are also significant technological stumbling blocks.

(a) _Space Heating and Cooling_. Space heating with hot water is not new. Many American homes are currently heated by hot water circulated from a fossil-fuel furnace. In England, the waste heat from power generation has already been substituted for individual furnaces in some communities, and the famous community of Tapiola in Finland with 20,000 inhabitants has been using waste heat as a heat source since 1953.

The critical time for thermal pollution, though, is summer, not winter. Where would the waste heat go in the summer? Some engineers have proposed a dual system that would use an absorption refrigeration process with ammonia or lithium bromide refrigerant to shift the waste energy from supplying heat during winter to supplying air conditioning in summer. The peak heating and cooling requirements for a typical three-bedroom house are almost matched at 75,000 BTUs per hour. Yet these proposals still do not indicate how heat can be discharged in the fall, spring, or other periods when the weather does not require house heating or cooling.

(b) <u>Aquatic Farming</u>. Aquatic farming using the effluent from power stations has already been established as economically feasible. Yet the problem of disposing of heat during the summer typically remains. Thus, the deliberate use of steam plant effluents in aquatic farming has been limited to situations where the receiving water is either continually cold, as in New England estuaries, or where the flow can be switched to a less sensitive discharge point during the summer, as would be possible from some coastal generating plants.

Some examples are already available. The Long Island Lighting Company has been experimenting with the growing of oysters in the discharge from their nuclear power plant at Northport, New York. Some experiments have been conducted with lobsters in Maine, and the raising of shrimp in the effluent of the Turkey Point power plant in Florida has been studied. In England, there have been proposals to raise ornamental ("tropical") fish in effluents.

(c) <u>Agriculture Use</u>. Agricultural losses from frozen crops mount into the thousands, sometimes millions, of dollars annually. The logical idea has occurred to a number of people that thermal discharges could be diverted in high-value agricultural areas, such as citrus groves, and losses could be cut. Agricultural lands, though, are typically distant from urban customers for energy, sources of coal or fuel oil, or any of the other typical justifications for power-plant location. And the need for protection of crops from freezing occurs only a few days each winter, hardly the justification needed for the cost of distributing vast quantities of water continually pouring from a power plant every day of the year.

(d) <u>Water Conditioning</u>. Experiments have shown that flocculation, the precipitation of impurities with chemicals in municipal water treatment, increases in efficiency and effectiveness with raised temperatures. In 1962, the State of Pennsylvania's Committee on the Effects of Heated Discharges estimated that fifty cents per million gallons of water treated could be saved in the cost of chemicals for each 10° F. rise in temperature. Water treatment typically ranges in total cost from three to thirty dollars per million gallons. While a saving in the vicinity of fifty cents or one dollar may be advantageous in some cases, rarely would a water treatment plant or power plant be located on this justification alone.

(e) <u>All-Season Shipping Lanes</u>. Shipping channels in northern United States, notably the St. Lawrence Seaway, are closed each winter because of ice blocking the channels. Industries have to pay for higher transportation rates, and ship-related services have to be suspended. A hope is being expressed by some observers that effluents from nuclear-reactor power plants can be used to keep the channels clear of ice through the entire year. For example, a 600-megawatt reactor plant is estimated to generate enough waste heat to keep an eleven to sixteen-mile reach of the St. Lawrence River free of ice. Yet no detailed consideration has been given to the effect of these plants upon the river's biota. What price should we pay for several extra months of ice-free river traffic?

(f) <u>Heat Balancing in Rivers</u>. A totally different frame of reference has been used to suggest that we use waste heat to balance the effects that dam construction has had on the heat levels of some rivers. Water in a rapidly flowing river mixes and has a fairly constant temperature that is sensitive to its thermal and energy surroundings. When water is impounded behind a dam, though, thermal layers similar to those in any deep lake form. And water drawn from the reservoir has temperature characteristics that bear little resemblance to those of the original river. During the biologically active summer, especially, the water is likely to be considerably colder since it is often drawn from the bottom of reservoirs. So some observers have suggested that thermal discharges be mixed with the reservoir discharges to attain a more appropriate summer balance. In the winter, the combined discharges would be warmer than the original flows, but the biota would be more quiescent and the environment better adapted to absorbing the heat.

Cooling Technologies

When viewed critically, it should be apparent that few of the suggested useful applications for waste heat are really suitable for the heat discharges occurring from power plants. In most cases, the heat need is intermittent, but the supply is not. Costs would typically be so high that resources could be better appropriated on other social needs. Ecosystem effects have not been systematically studied. Thus, we have to turn to technologies to harmlessly dissipate the heat. This is not difficult. We have the technologies; but, unlike Europe, we have been unwilling to pay the relatively marginal cost.

In the environment, only one practical sink for waste heat exists--the atmosphere. Technically, the problem is to transfer the heat from water to air. Currently, three alternatives are being used: (1) once-through cooling using surface transfer from natural rivers, lakes, or estuaries; (2) cooling towers; and (3) cooling ponds. In 1955, only about one out of six plants used towers or ponds, and most of these were in the arid Southwest. Also, they were typically small plants with a capacity less than 400 megawatts. But now new plants are being planned with capacities of more than 4,000 megawatts, and the quantity of cooling water-- more than 21,000 cubic feet per second in the case of a nuclear reactor--needed to limit the temperature rise of the river to 5° F. would exceed the average flows of many rivers in the Northeast.

Once-Through Cooling. Engineers using once-through cooling (removing, using, and then immediately discharging the cooling water back into the natural source) have a choice of two basic strategies: (a) forming a heat layer, either surface or sub-surface, or (b) mixing the heated water with the natural. Each has advantages and disadvantages, and often several approaches will frequently be used simultaneously. If the heated water is simply "floated" on top of the natural, heat dissipation into the atmosphere will be the most rapid and the stream will typically return quickly to its original temperature. But the layer of heated water will form a blanket over the natural, severing oxygen transfer and having the most severe impact upon the biologically richest strata--the top few centimeters--of any natural waters. Heated water released below the surface may be less harmful, but the heat cannot be dissipated as rapidly, and it will divide the aquatic ecosystem like the modern expressway severs a city.

The alternative is to use mixing jets, diluting the heated water with the cooler natural water. But this is like avoiding the scorching of the first half-mile to par-boil several miles. Dissipation of the heat into the atmosphere is delayed, and effects will be measurable further downstream. When designing the jets, both the local aquatic biota and the migratory biota must be considered. In the Northwest, it was found that the plume of heated water frequently deflected and, in at least one case, blocked the migration of salmon.

Cooling Towers. An engineer designing a cooling tower to dissipate waste heat faces three sets of decisions: (1) His tower can be either "wet," the water being exposed directly to the air, or "dry," like the water being circulated in an automobile radiator. (2) The air entering the tower may use a "natural draft," the natural tendency of heated air to rise, or rely upon "mechanical draft" (fans) to push or pull the air through the tower. (3) And the airflow can be either horizontal (a cross-flow design) or vertical (a counter-flow design) through the heat-transfer section of the tower.

Wet towers with natural draft and either counter-flow or cross-flow are now proving most popular. Before 1963, only mechanical draft towers had been used in the United States. By 1968, over twenty natural draft towers were in various stages of construction or in operation. The largest was almost 450 feet high and over 300 feet in diameter at its base. Even a modest nuclear-fueled plant of about 400 megawatts would require a vertical flow tower about 400 feet high and 285 feet in diameter. If a counter-flow design is used, the height would drop to 370 feet, but the diameter would increase to 380 feet.

Mechanical towers have fallen from favor because of their high maintenance and operating costs, which are annually often several percent of the original capital cost of the tower. Federal authorities have estimated that a typical evaporative cooling system will increase the capital cost of a power plant by about 5%. But it is also noted that this increase is not carried to consumer costs since the cost of transmission and administration represent about four-fifths of your electric bill. Hence, the use of cooling towers and the recirculation of the cooling water, meaning virtually the elimination of thermal pollution, would raise your electric bill by about 1%.

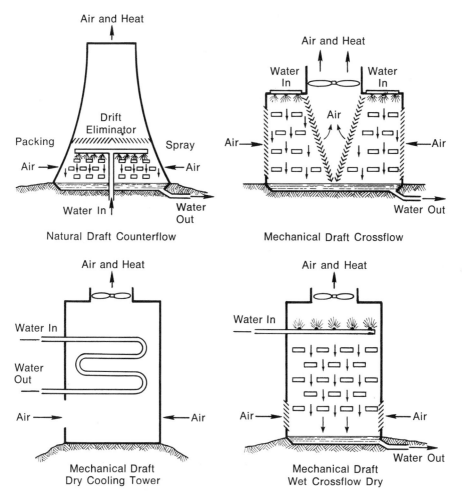

Figure 7-4. Alternative cooling tower concepts.

In the early days of cooling tower construction, faulty design and poor siting caused some problems of fog or mist formation in the areas around the towers. This is now considered a negligible problem because of the improved designs and more intelligent selection of sites, especially if a highway is nearby. Some water will be lost, however, through evaporation, and this will have to be replaced by more water from a natural source.

Cooling Ponds. Flooding land to create a reservoir of cooling water that is recirculated represents the simplest and, occasionally, the least expensive means of eliminating thermal pollution in natural ecosystems. The warm water is discharged into one end and, eventually, withdrawn as cold water at the opposite end of the reservoir. Since cooling will be slow, about two acres per megawatt capacity of the plant is usually needed if an ecologically stable and attractive pond environment is to be maintained. For a 1,000 megawatt plant, this would mean a pond area--and land purchase--of 1,000 acres.

Pond size can be decreased by spraying the water into the receiving end, but use of a cooling pond becomes prohibitively expensive if land costs are high. And fog tends to form on cold days, extending as far as one-fifth mile downwind from the pond. Aesthetically, and ecologically, the choice between the cooling pond and cooling towers would depend upon local circumstances; but, if sensitively designed, a cooling pond can be a recreational and scenic asset.

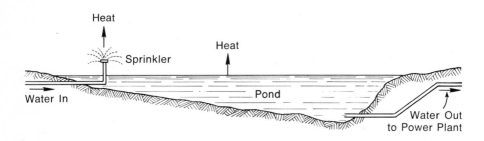

Figure 7-5. Profile of cooling pond with spray.

Conclusion

Unlimited discharge of waste heat into the environment cannot be tolerated any longer. The most liberal criteria today for releases of waste heat recommend a maximum water temperature rise in the stream of 5° F., and the maximum rise may be as low as 1% in some estuarine and lake environments. Inevitably, all power generating plants and some industries will be forced to spend funds for cooling technologies. Given the relatively small significance of this cost to most of the final products, including electric power, the amazing aspect of the entire problem is the exorbitant length of time that thermal pollution has been allowed to occur in the United States.

QUESTIONS FOR DISCUSSION

1. Using a nearby river as a case example, describe the biological effects that you would expect from a thermal rise of 5° F. in the stream temperature.

2. Describe the confluence of climatic, river discharge, and thermal pollution conditions that you would expect to prove most lethal to aquatic life.

3. What changes in the regulation and ownership of electric power utilities might encourage more constructive use of waste heat?

4. Explain the difference in biological effects between thermal pollution and oxygen depletion by sewage in an aquatic environment.

5. Should the heat and water-vapor discharges from a wet cooling tower be considered pollution?

SUGGESTED READINGS

Two excellent surveys of the problems and potential solutions to thermal pollution exist. The first is written in a less technical language. The second contains a particularly helpful bibliography on the engineering aspects.

1. U.S. Department of the Interior, Federal Water Pollution Control Administration, Industrial Waste Guide on Thermal Pollution (Corvallis, Oregon: FWPCA, Pacific Northwest Water Laboratory, 1968).

2. Frank L. Parker and P. A. Krenkel, Physical and Engineering Aspects of Thermal Pollution (Cleveland, Ohio: CRC Press, 1970).

In June, 1970, a symposium on "Thermal Considerations in the Production of Electric Power" was held in Washington, D.C., under the sponsorship of the Atomic Industrial Forum's Committee on Environmental Law and Technology and the Electric Power Council on Environment. While most of the papers in the published volume that followed the symposium are superficial and represent the power industry's point of view, a few thoughtful and informative papers can be found. See:

3. Merril Eisenbud and George Gleason (eds.), Electric Power and Thermal Discharges: Thermal Considerations in the Production of Electric Power (New York: Gordon and Breach, 1969).

An interesting case study that illustrates the social, political, and technological complexities represented by power-plant construction and, possibly, thermal pollution in a deep-water lake has been prepared by Cornell University's Science, Technology, and Society Series. See:

4. Dorothy Nelkin, Nuclear Power and Its Critics: The Cayuga Lake Controversy (Ithaca, N.Y.: Cornell University Press, 1971).

An immense number of books have emerged in the past few years evaluating energy sources. Examples would be:

5. Allen L. Hammond, W. D. Metz, and T. H. Maugh, II, Energy and the Future (Washington, D.C.: AAAS, 1973).

6. Committee on Power Plant Siting, Engineering for Resolution of the Energy-Environment Dilemma (Washington, D.C.: National Academy of Engineering, 1972).

ESSAY

Peak Power Requirements and the
Pumped Storage Controversy

The Rise in Pumped Storage

Pumped storage projects have suddenly riled conserva-
tionists throughout the United States. If electric power
utilities have their way, the irritation will continue.
From 1.1% of the installed generating capacity in 1970,
pumped storage generating plants are expected to increase
their share to 5.5% by 1990--an increase in absolute capacity of twenty times. The battles will be noisy and bloody
for both reputations and feelings. Consolidated Edison of
New York, for example, has doggedly fought for over a decade
to flood the peaks of Storm King Mountain above the Hudson
River near New York City despite myriad law suits and extremely voracious criticisms. But Consolidated Edison, like
many other utilities, has considered the pumped storage
sufficiently inexpensive and convenient compared to alternatives that the battles have seemed worthwhile.

Base and Peak Loads

Pumped storage--the pumping and subsequent release of
water to and from elevated hydroelectric reservoirs--is a
product of (a) our erratic use of electric energy and (b)
the inflexibility of the immense steam-electric generating
plants. During a typical summer day, an urban community
will be using a modest amount of electricity during the
early morning hours. Use rises a bit when sleepy citizens
turn on electric toasters, open refrigerator doors, and
begin cooking breakfast on electric stoves. When they arrive at their offices and switch on air conditioners, electricity demands skyrocket. In New York City during mid-July
of 1969, for example, demand doubled during a hot weekday
from about 3,500 megawatts at 5 a.m. to almost 7,500 by 5
p.m. Then use dropped sharply. In some cities, the rise
may be even more drastic, peaking at perhaps three times
the minimum, or "base," demand. During winters, the peaks
occur earlier and later due to use of electric lights, and
there may be a dip in use during the middle of the day.

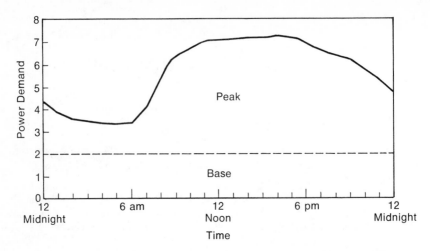

Figure 7-6.   Typical daily power demand fluctuation.

Steam Generation Inflexible

Combustion levels in steam boilers or the steam temperatures and pressures cannot be changed suddenly to cope
with erratic fluctuations in demand. And even minor changes
in steam-energy levels cause inefficiencies. Utilities
prefer to operate their steam generators for the base load,
the amounts that are used consistently; and the steam produced during lapses of demand is simply wasted.

By contrast, hydroelectric generation is quite flex-
ible. Flows can be stopped and started at will, which is
ideal for peak loads. But the quantity of water behind a
hydroelectric dam is usually limited, so engineers prefer
not to use it for the base load. This is especially true
where the reservoir is used for recreation and the public
prefers to have a constant lake level without expanses of
mud flats.

So steam and hydroelectric generation comfortably com-
plemented each other as long as hydroelectric projects could
be built to keep up with peak demand. But hydroelectric
sites have always been limited in the Northeast and are becoming scarce elsewhere. Alternative techniques for peak
power generation are needed. Gas turbines can be used, but
in their current state of development they are noisy and
relatively inefficient, having a thermal efficiency of about
25%.

162

Pumped Storage Attractions and Problems

In the early sixties, utility engineers decided that their solution to the peak-load problem would be to create a new type of hydroelectric facility that would not depend upon river reservoirs. Hilltops are rarely developed, and the land can be purchased inexpensively. If a few undulations in the hilltops could be enhanced by levees, a reservoir could be formed. Then water could be pumped into the reservoir from a nearby river or lake during the night when steam plants are normally wasting energy. Water would be released during the day to drive generators and fill the peak demand.

For utilities with the single-purpose objective of providing electric energy by the cheapest means, pumped storage is a marvelous concept. But there are environmental problems. Wooded hilltops near urban centers typically represent favored recreational resources. Watching water recede across desolate mud flats towards a gaping hole is not the tranquil environment favored for picnics. Nor is this up-and-down water suitable for aquatic life. Sensitive wildlife that might be sucked into the pumps on the river or lake would be killed by the pressures and turbulence. And the pipes cutting swaths up and down the hills do not enhance scenic beauty. Nor are the massive transmission lines strung over the landscape particularly decorative.

Environmental Lessons Needed

Today power companies are continuing to lobby for more pumped storage, and they have been remarkably successful in this quest when the environmental damage is considered. But alternative technologies do exist--albeit slightly more expensive--and power companies will undoubtedly be forced to turn towards gas turbines and special "topping" technologies that are in varying stages of development.

Chapter 8

SIGNIFICANCE OF AIR POLLUTION

> "It is this horrid smoake which
> obscures our churches and makes
> our palaces look old, which
> fouls cloth and corrupts the
> waters . . . [and is] so uni-
> versally mixed with the other-
> wise wholesome and excellent
> aer, that [London] inhabitants
> breathe nothing but an impure
> and thick mist, accompanied
> with a fuliginous and filthy
> vapour, which renders them ob-
> noxious to a thousand inconven-
> iences, corrupting the lungs,
> and disordering the entire
> habit of their bodies; so that
> catharrs, phthisicks, coughs
> and consumption rage more in
> this one city than in the whole
> earth besides." John Evelyn
> (1661)

The Nature of Air Pollution

Definition

Air pollution is the presence of "foreign" materials in
the air. Everything that can be vaporized or made into
small enough particles to float in the air should be classified a potential pollutant. Air is considered normal if
more than 99.99% of the air volume consists of only four gas
molecules, nitrogen (approximately 78.09%), oxygen (20.94%),
argon (0.95%), and carbon dioxide (0.03%), and about a dozen
other constituents are found in trace quantities usually expressed in parts per million. But the typical urban air
will contain some of these trace materials, including carbon
monoxide, sulphur dioxide, and methane, in excessive amounts.
And, in addition, there will be measurable levels of mercury, cadmium, zinc, lead, chlorine compounds, carbon,
silica, various hydrocarbons (rubber, oils, methane,

propane, plastics, and other formulations), asbestos, and a chemist's cacophony of industrial compounds and their derivatives. These are sometimes referred to as primary pollutants since human technology is usually directly responsible while still other compounds are known as secondary pollutants because they may be formed when the man-made emissions combine--usually aided by sunlight providing energy--with oxygen, nitrogen, and water vapor in the air. Ozone (O_3) and peroxyacetyl nitrate (PAN) are among the more common.

There is also a pollution that can be classed as natural. Many gases, including hydrocarbons and sulphur oxides, are formed in the biological decomposition of materials. Pollens from trees and weeds or spores from fungi can cause acute misery to humans. Odors, which are caused by trace molecules in the air, can be classed as either natural or man-made pollutants.

The Standard Pollutants

In an urban setting, air pollution discussed in official documents usually refers to five groups of materials plus their secondary products. These are the substances commonly measured by air-pollution control agencies, and the levels are duly relayed to the federal government for national compilation. And they have been implicated in most--but not all--of the destructive effects of air pollution. All have common urban sources (see Table 8-1).

Carbon monoxide (CO) is formed when fossil fuels do not burn completely. Automobiles represent the overwhelming source, and formation of carbon monoxide could be largely prevented by more oxygen and higher engine temperatures--but these conditions usually produce more nitrogen oxides. Carbon monoxide is colorless, odorless, poisonous, and slightly lighter than air. It is assumed to eventually convert to carbon dioxide, but the mechanisms are not thoroughly understood.

Particulates are the solid and liquid particles that float in the air. Cinders and soot from the combustion of coal or oil for electrical energy generation and incineration of wastes are the most common sources, but the burning, grinding, and abrasion of anything solid or splashing of a liquid typically produces particulates. The size may vary from visible soot and smoke to particles detectable only under an electron microscope. They may drift to the earth immediately after leaving a chimney, or they may be carried for thousands of miles by air currents.

Sulphur oxides traditionally have been emitted from burning of coal and oil in electrical-energy generation or space heating. Sulphur is naturally found in many coals and oils, and widespread legislation encouraging the use of low-sulphur types of coal or oil or control devices has significantly improved air quality over recent years. The gas is acrid, corrosive, and toxic, but the primary threat to health occurs when the sulphur dioxide combines in the air with water vapor and other compounds to form sulphuric acid and sulfates. Being a relatively heavy gas, sulphur dioxide hugs the city streets; and, except when emitted from exceptionally high smoke stacks, most of the gas will have fallen from the air within several hundred miles of a major metropolis.

Hydrocarbons are usually the unburned fumes that evaporate from gas tanks and are emitted from exhausts of vehicles. But they can also be the evaporating solvents of asphalt, gaseous emissions of rotting vegetation, or the product of any reaction that involves an organic (carbon-containing) material. While they are not considered toxic in normally found concentrations, they have been implicated as causative or contributory agents for some cancers. They also contribute to the dark, yellow haze that shrouds cities.

Nitrogen oxides are produced whenever air is heated to high temperatures, such as in an automobile cylinder or in the high-temperature furnace of a power plant. Usually inert atmospheric nitrogen combines with the oxygen to form nitric oxide (NO) and, later, this converts to nitrogen dioxide (NO_2). The gas has a yellow-brown color, and it is mildly irritating to lungs in low concentrations. When combined with rain, nitric acid is formed.

Sources

Making a rather crude estimate, the Environmental Protection Agency has reported that transportation sources, especially automobiles, are responsible for over three quarters of the carbon monoxide and over half of the hydrocarbons and nitrogen oxides released to the nation's atmosphere (see Table 8-1). Particulates are primarily products of industrial processes, and most sulphur dioxide emissions can be attributed to the coal combustion by electric

166

generating plants. These figures, however, can be mislead-
ing to anyone facing a local air-pollution problem. Every
city will vary in its particular mixture and concentration
of air pollutants. Considerable heavy industry will typically mean large quantities of particulates. A residential
community with streets crowded by automobiles can expect
high levels of carbon monoxide, hydrocarbons, and nitrogen
oxides. A city in a mountainous area gripped by winter
weather and lacking low-sulphur fuels can anticipate sulphur
dioxide problems, while another city open to breezes may
rarely experience any difficulties with air quality.

Table 8-1. Estimated Emissions of Air Pollutants
by Weight, Nationwide, 1971
(In millions of tons per year)

Source	CO	Partic-ulates	SO_x	HC	NO_x
Transportation	77.5	1.0	1.0	14.9	11.2
Fuel combustion in stationary sources	1.0	6.5	26.5	.6	10.2
Industrial processes	11.4	13.6	5.1	5.6	.2
Solid waste disposal	3.8	.7	.1	1.0	.2
Miscellaneous	6.5	5.2	.1	5.0	.2
Total	100.2	27.0	32.6	26.6	22.0
Percent change 1970-1971	-.5	+5.9	-2.4	-2.6	0

Source: Environmental Protection Agency.

It is now obvious that improvements will occur slowly,
if at all, because sources are closely related to desired
styles of life and economic interests. Concerted efforts
have been made to regulate major air-pollution sources, such
as the automobile, and some progress, especially in automobile emissions, has been registered in the last few years.
But other pollution in other categories, particularly particulates, has increased in many cases. Industry continues

to expand. More automobiles are being purchased and driven, eroding air quality gains. Because of air-pollution control devices, automobile efficiency has been reduced, increasing the potential for pollution. And the shortage of petroleum has placed a severe burden on alternative low-pollution sources of energy. Thus, revised air-pollution programs and new control approaches can be expected to be a continuing feature of the political scene.

Objections to Air Pollution

Air pollution today is pervasive and relatively unnoticed. Unlike a century ago, we do not have thick black smoke of locomotives and factories constantly clouding the sky or the choking fumes of sulphur dioxide spilling from every chimney. Most gases produced by modern technology are colorless and relatively odorless. But they still exact a toll. Air must be breathed, and pollutants affect health. Property is corroded. The natural ecosystem is crippled. Aesthetic senses are affronted. These objections are frequently not dramatic, and there is a tendency by communities to postpone action on any problem that is less than obviously acute, especially when the action may be inconvenient or expensive.

Health

Describing the direct physiological effects of air pollution on humans can be like describing the shape of an iceberg floating in the sea. A few historical episodes, periods of intense air-pollution concentrations when mortality rates have risen sharply, stand out clearly and are described thoroughly in the literature. Numbers of persons, symptoms, circumstances, and the associated statistics can be clearly linked to air pollution without strong dissension. But we know that the most massive health problem with air pollution is not associated with identifiable episodes but in the gradual erosion of health by frequent and long-term exposures. Hypotheses linking this type of exposure with specific illness require murky assumptions and estimates that are easily attacked piecemeal by dissenters.

By extrapolating from industrial situations, laboratory experiments, and a few epidemiological studies, however, we can relate a few health conditions with several common pollutants. Even general predictions can be made of the health response to different levels of some specific pollutants over short periods. Yet, except in a few cases, such as asbestos, we do not typically know the long-term--twenty-five or thirty years, for example--responses to low levels of exposure. Nor do we know how to predict the health impact in a typical urban situation when an individual may be engulfed by dozens of air pollutants simultaneously. We can suspect that they will act synergistically, the combined danger being greater than each individual danger added separately, but we rarely can find "conclusive" evidence.

Episodes. In describing the health hazards of air pollution, the technical literature invariably cites a set of historical episodes. These provide dramatic examples of air pollution effects but, as noted earlier, should not be interpreted as indicating total effects. For three days in early December, 1930, a thick mist covered Belgium with an especially dense concentration along the narrow River Meuse Valley, an area not longer than 12 miles and wider than 1.5 miles. Air was still and cold, temperatures during the day being around freezing. Smoke from numerous factories mixed with the fog to produce a choking smog. Residents became hoarse, short of breath, and frequently nauseated. Sixty deaths were attributed to acute heart failure caused by the air conditions. Most were elderly people with weakened hearts and lungs, but even the young and healthy became seriously ill. Cattle also showed symptoms, and some eventually died.

Similar circumstances existed in the horseshoe-shaped Monongahela Valley around Donora, Pennsylvania, in October, 1948. Toxic fumes and smoke were rising from numerous factories along the river, and freight trains operated on both banks with coal-burning, steam locomotives. The air was cold and damp. For more than five days after October 26th, no breezes disturbed the thickening smog that accumulated in the valley basin. About twenty deaths were attributed to the episode, and nearly 6,000 persons--representing nearly half of the area's population--were stricken with irritation of the eyes, nose, and throat, labored breathing, coughing, chest pains, headaches, nausea, and vomiting.

The topography of London was different but other circumstances were familiar. On December 4, 1952, a mass of cold, damp air enveloped the lower Thames estuary. The fireplaces of London buildings were lit, and the smog gradually became so dense that traffic became stalled. About

160 prize cattle in a show at Earl's Court developed breathing difficulties and fever.

Only after the fog had lifted did statisticians and the population realize the extent of human casualties. For the week ending December 13th, 2,851 more persons died than would be expected under normal circumstances. During the following weeks, an estimated 1,224 further deaths were attributed to the smog. In 1956, another smog lasting eighteen hours produced about 1,000 deaths above the usual rate. Other publicized episodes have occurred in New York, New Orleans, Yokohama, and Los Angeles. And normal winter conditions in some cities, such as Ankara and Seoul, in the less developed countries where only low-quality fuels are available can be likened to a constant episode though reliable mortality and morbidity statistics are lacking.

<u>Erosion of Health</u>. Most victims of air pollution will not die during an air episode. They will contract a respiratory disease or another symptom associated with air pollution, gradually weaken, and then typically die from pneumonia, a heart attack, or failure of some other vital organ. Or they will bear a child with a birth defect that future medical research will link to an air pollutant. Or perhaps they will develop a disease, such as cancer, caused by a dimly understood set of factors with air pollution as only one possible component.

Air pollutants affect health by three principal mechanisms.

(1) They alter enzyme action. Enzymes are complex proteins that govern many metabolic activities by intricate and vulnerable interactions. Arsenic and mercury, for example, have an affinity for certain enzymes and block them from further effectiveness. A pollutant that reacts with manganese, on the other hand, may inactivate a manganese-dependent enzyme indirectly by removing the manganese. Other pollutants may simply destroy the enzyme, as occurs when an organophosphate pesticide attacks the enzyme that regulates muscular responses. In still other cases, the pollutant may stimulate the enzyme into producing a toxic substance.

(2) Pollutants may react directly with cell constituents. Carbon monoxide, for example, displaces oxygen in the blood's hemoglobin, thus starving the brain and other tissues, of the oxygen.

170

(3) Pollutants can influence the chemical activities of the body. Liver damage from carbon tetrachloride poisoning, for example, is believed caused by the discharge of epinephrine by nerves that have been stimulated.

The precise effect of air pollution cannot usually be predicted, however, because a person is normally exposed to mixtures of air pollutants. Some will intensify a reaction when combined while others may counteract each other, reducing the effect. Factors influencing the human response may include the person's age, health, and physical stress as well as the length and dose of exposure.

Defenses. Not all pollutants that are breathed affect health since many are removed safely by the body's defenses. Soluble gases will be absorbed in the trachea (wind pipe). Particulates larger than ten microns in diameter will be filtered by nose hairs or held by the moist lining of the nose, and particulates larger than two or three microns will usually be removed before leaving the trachea and entering the lung. But particles smaller than several microns can be carried past these defenses into the lung tissue where they can be absorbed by the body. Furthermore, these particulates frequently have gas molecules, such as sulphur dioxide, adsorbed to their surfaces, providing a more dangerous pollutant than the particulate alone.

In summary, human health is the paramount but still elusive justification for strong control programs of air pollution. Intuitively, the belief exists among health statisticians that air pollution is a significant factor in diseases and deaths within industrial society. But conclusive evidence that specific levels of improvement in air quality will have a predictable effect on human health is still lacking. Statements in a few recent reports, such as the National Academy of Sciences/National Research Council findings in 1972 that lung cancer is twice as common among city dwellers as rural residents and is more common in areas of cities where general industrial pollution is worse, approach an unequivocal conclusion. Yet, until a concensus similar to the one against smoking has crystallized, use of health considerations in arguing against air pollution will remain an invitation to controversy.

Property Damage

About thirty-six studies on the costs of air pollution damage have been conducted, and these were summarized in 1973 by the Environmental Protection Agency's report, Cost of Air Pollution Damage: A Status Report. For 1968, the national cost of air pollution damage was estimated to total $16.1 billion which included $5.2 billion for residential property, $4.7 billion for materials, $6.1 for health, and $0.1 billion for vegetation. The cost of damage from soiling was excluded from this estimate to avoid counting twice the cost of damage to property and materials.

In more personal terms, this cost represented about $80 per capita or $240 per family. But these gross figures disguised the cost for residents of major cities. For example, a United States Public Health Service estimate in 1968 indicated that Manhattan residents were paying an annual average extra cost of $220 a year for cleaning, homes, and vehicles soiled by particulates. Dollar estimates of air-pollution damage can be misleading, however, because of the vast variety of materials being affected.

Metal. Sulphur dioxide in the air has been incriminated as the primary cause for decay of copper roofs, zinc coatings, steel plates, and the contacts in electrical equipment. Moisture is critical to corrosion because, in the presence of an acid pollutant, it forms a crude electric cell that allows current to flow and a chemical change to occur. Below 60% relative humidity, corrosion is slow while there is a marked increase above 80% relative humidity. During periods of rain, corrosion is considered minimal.

Building Stone. Sulphur dioxide has also been identified as the principal source of the acid that dissolves carbonate (lime) stones in an urban environment. But the black particulate matter that coats the stones accelerates the process by providing a spongy film for holding the sulphur dioxide against the stone surface. Egyptian monuments in London and New York have corroded more in the last few decades than in the previous three millenia. Stone figures on the cathedrals in Europe have been rapidly crumbling, and decaying facades of buildings in major U.S. cities have become safety hazards for pedestrians and passing vehicles below.

<u>Clothing</u>. Air pollution affects clothing in three ways.

(1) Fabrics are weakened directly. Nylon stockings
    disintegrate minutes after exposure to sulphuric
    acid aerosol. Other clothing will have useful
    life drastically reduced.

(2) Air pollution weakens fabric by soiling, causing
    more frequent cleanings.

(3) Colors are altered directly and indirectly.
    Dyes can be faded by some pollutants, and this
    tendency would be accelerated by more frequent
    washings. Also, particulates can stain cloth,
    changing the original color.

<u>Paint</u>. Air pollution reduces the life of protective
coatings and soils surfaces. Sulphides, for example, will
react with lead-based paints causing discoloration. Also,
sulphur dioxide will increase the drying time of lead-based
paints, thus allowing more particulates to settle on the
surface.

<u>Other Materials</u>. Unless a material is chemically inert
it can be damaged by air pollution. Books in libraries have
been weakened and discolored by sulphur dioxide, and colors
in museum paintings have been altered and darkened. Rubber
cracks when exposed to atmospheric ozone, and the accumulation of particulates on city streets and sidewalks increases
accidents, especially after a light shower.

Indirectly, air pollution exacts a cost by lowering
visibility, increasing the accident rate. Turnpike accidents in smog are commonplace. Between fifteen and twenty
airplane crashes in 1964 were attributed by the Civil
Aeronautics Board to air pollution limiting visibility.
Adding to the risk of accidents is the effect of carbon
monoxide in dulling consciousness and slowing reaction time
to emergency situations.

<u>Natural Ecosystem</u>. Airborne pollutants damage plants
in predictable ways. Plants normally must respire, transform light energy to chemical energy, and amass most of the
chemical constituents through a root system. But air pollution enters the leaf structure and destroys cells, blocks
the passage of light energy into the plant, and adds new
chemicals to the soil.

173

Wildlife typically is less sensitive than the plants, but like man, animals can be affected directly or indirectly. On a longer time basis, any changes caused by air pollution to plant life will eventually be reflected in the kind and abundance of wildlife.

Sensitivity of plants to air pollutants varies widely among species. Tobacco plants, for example, are known for sensitivity to ozone, and dahlias are reliable monitors of sulphur dioxide levels. Gladiolus and tulips can be used to identify airborne fluorides, and the pinto bean plant responds to the secondary pollutant, peroxyacetyl nitrate (PAN). Tobacco plants readily register damage from ozone, and petunias can serve as an indicator of any oxidants, such as ozone. But each plant maintains some individuality in its reaction. Amounts of available water, soil conditions, temperature, light intensity, fertilizer added, and many other conditions determine response.

The situation is complicated by the bacteria, viruses, insects, fungi, and climatic conditions that can cause damage similar to those expected from air pollutants. In general, though, each major pollutant causes a characteristic type of damage. Sulphur dioxide tends to cause blotches of damage that concentrates between the veins, appearing as dead areas. The needles of conifers (evergreens) become reddish on the edges as tissue dies and contracts. Ozone can cause a silvery bronzing of the leaf surface, and small speckles of damage cover broad areas of the leaves. Fluorides, which are usually highly toxic to plants, cause damage that will usually first appear in the tip or edge of the leaf with a distinct reddish-brown lime dividing the healthy cells from the dying. Particulates gradually coat the leaf's surface with a black film.

Sulfur Dioxide   Ozone   Fluoride

Figure 8-1. Leaf damage characteristic of
(a) sulphur dioxide, (b) ozone,
and (c) fluoride.

174

The extent of damage can vary widely. Spectacular examples can be found around some ore smelters. The landscape at Copperhill, Tennessee, has for decades provided a shock for travelers as they leave the luxuriant forests of the Smoky Mountains for the barren, eroded hills of the Copper Basin where a smelter operated in the 19th Century.

A more subtle manifestation of damage is a shift in the dominant species of a region. Ponderosa pines are reportedly declining in numbers wherever they are exposed to urban smog because of their sensitivity to ozone. In one area, the San Bernardino National Forest near Los Angeles, about 1,000 acres of Ponderosa pines had to be cleared in 1970 when an entire stand was killed by the ozone-laden smog from the San Bernardino Valley. On the East coast, the white pine is also especially susceptible to air pollution, and research efforts are being made to develop a more resistant variety. In general, conifers rarely thrive in cities because the air pollution causes excessive shedding of needles. And truck gardeners no longer attempt to grow some crops, such as spinach, near cities.

At the same time, trees and other vegetation remain one means of improving air quality. Trees are noted for their ability to "scrub" pollutants from the air, physically stopping their movement or chemically absorbing them. Trees also alter the air circulation at lower levels, and groves can be used to prevent the massing of numerous air-pollution sources.

Climate Modification

Our lives are intimately linked to the existing climate pattern, and the threat of significant shifts in climate caused by air pollution has been one of the more effective specters raised by scientific soothsayers of the past decades. On a global scale, pollution from human activities is gradually beginning to approach proportions that could be significant though the factors are extraordinarily complex. On a regional scale, though, we have already reached the point of affecting climate, and some evidence is available.

Four issues--the possible greenhouse effect from carbon dioxide, an altered albedo (reflectivity to light energy), artifically induced rainfall, and excessive thermal discharges--dominate the debates over inadvertent climate modification. The concern over a possible greenhouse effect arises from the scientific principle that short-wave light energy from the sun can pass relatively easily through carbon dioxide but the longer, infrared waves of radiant heat energy are absorbed. Glass has the same effect as carbon dioxide, so the principle is used in a greenhouse to capture heat energy beneath the greenhouse glass. The same phenomenon explains the build-up of heat in an automobile parked in the sun during the summer. In theory, the earth's atmosphere will become excessively hot if the carbon dioxide levels rise and capture more of the radiant heat being emitted from the earth's surface.

The validity of this concern depends upon two assumptions: (1) Carbon dioxide levels will continue to rise proportionate to our combustion of fossil fuels. (2) This rise will produce a simple, positive change in atmospheric temperature. Both assumptions are possible but neither are certain. We do not understand the mechanisms governing the level of carbon dioxide in the world. We know that vast amounts are stored in the oceans, but plants and bacteria also have the ability of storing carbon compounds. If carbon dioxide levels continued to rise, would these storage mechanisms remove more from the atmosphere to restore balance? And if the climate did begin to change, would corrective forces be stimulated to restore a cooler climate. Evidence is inconclusive. Combustion of fossil fuels is believed to have produced an increase in atmospheric carbon dioxide at the rate of 0.2% annually since 1958. An 18% increase in carbon dioxide levels is projected for the year 2000, and crude calculations suggest the surface temperature of the earth may rise 0.5% centigrade. A doubling of the carbon dioxide levels would raise temperatures by 2 degrees centigrade. This could lead to partial melting of the polar ice caps; and this, in turn, could precipitate further climate shifts.

Fine particles in the atmosphere can be even more effective than carbon dioxide in altering the earth's heat balance since they will both reflect light energy before it enters the atmosphere and absorb it during passage to the earth's surface and return. Vast numbers of natural particles, such as dust·, volcanic ash, sea spray, are already circulating. But mankind is adding increasing quantities by agriculture, deforestation, and combustion of fossil fuels. If these were sufficiently numerous, we suspect the overall effect would be to decrease the atmosphere's temperature, but we are not even certain of that.

Small particles can also act as nuclei for the condensation or freezing of water vapor in the air. Again, the
question of relative significance of man's contribution compared to the levels existing naturally must be asked. The
long-term implication of man's increasing release of heat
into the atmosphere is also troublesome since it could
theoretically upset the earth's heat balance. Production
of energy is now rising at a rate of about 6% annually, and
this could amount to as much as 1% of the net radiation
energy from the sun over the U.S. continent sometime in the
early 21st Century. This heat addition, though, would center in urban areas, creating complex alterations in atmospheric circulation. Again, our current tools of analysis
are insufficient for predicting the outcome.

Regional changes in climate are more easily identified,
but even these effects are ambiguous. Increased rainfall
around some major cities has been recorded, presumably because the dust and chemicals, including automobile exhaust
lead combining with iodine and bromine from industry, have
provided nuclei for condensation and freezing of water
vapor. One intensely studied example involved the town of
LaPorte, Indiana, about thirty miles east of the Gary-
Chicago industrial complex. Compared to other weather stations not exposed to winds from Gary, LaPorte had 31% more
precipitation, 38% more thunderstorms, and 246% more days
with hail between the years of 1951 and 1965. The trend
towards more rainfall began about 1921 and roughly conforms
to the increase in smoky and hazy days recorded in Chicago.
Furthermore, many of the thunderstorms occurred in the early
morning, suggesting that heat rising from the stones and
furnaces of Chicago during the cool nights was responsible.*

Yet any theory that projects constantly more rain with
increased industrialization and urbanization is challenged
by a contradictory theory that an overabundance of particles
acting as potential nuclei could compete for water vapor
and inhibit the tiny droplets from coalescing into rainfall.
Cloud cover from particles has also presumably increased the
albedo of the atmosphere, shielding the city from the heat
of solar energy and tending to lower temperatures. But heat
island phenomenon indicates that the opposite occurs, probably because of the heat being generated within the city,

---

*Cities are known to be "heat islands," with average temper-
atures higher than surrounding areas. As a result, the air
over cities tends to rise.

the increased absorbancy of solar energy by urban surfaces, and the absorption of radiant heat by the particles themselves. In effect, scientists are still trying to sort out the various counteracting forces, and more time will have to pass before we can predict the climate effects of air pollutants. Meanwhile, only the foolhardy would disregard the potentialities.

Factors Affecting Air Pollution

People and Nature

Factors determining the severity of air pollution can generally be classified under two broad categories, one associated with human activities and the other with the natural environment. People design the technologies and choose the energy sources that lead to polluting emissions. These aspects will be the subject of the next chapter. But there is also a set of natural factors that influence the location and seriousness of a pollution problem. The most important are meteorological (related to the atmosphere and its phenomena, especially weather) and topographical.

Meteorological

To prevent a disastrous accumulation of pollutants, we depend upon air movements to dilute the gases and particulates and, eventually, to facilitate their removal by fall out, wash out, and chemical reactions. Three factors are particularly important: temperature gradient, wind, and precipitation.

Temperature Gradient. The desirable meteorological condition for dispersal of air pollutants is instability because gases should rise, expand, and become diluted and scattered. For this condition to exist, temperature in the lower atmosphere (troposphere) must become continually cooler as it rises in elevation. (See Figure 8-2.)

Occasionally, however, stable conditions that suppress this vertical movement will occur. This is an inversion, a condition where a layer of warm air exists above cool air. (See Figure 8-3.) When gases are released in the cool air, they rise until they reach the warm layer and then stop. Of the three common types of inversions, only one is considered

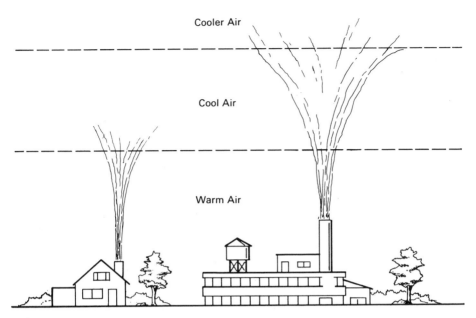

Figure 8-2.   A normal atmospheric gradient exists when the
air becomes cooler with elevation. (From
McCabe and Mines, Man and Environment, Vol. I,
Prentice-Hall, Inc., 1973.)

serious. A usually harmless type is the radiation inversion
of early mornings in a rural setting. The earth is cool,
but the air a few feet above the ground is still warm.
Mist will form along the ground, in hollows, and over ponds.
But the rising sun will heat the earth's surface, restoring
the unstable condition so the mist fades.

Another inversion condition is the frontal inversion
occurring when two air masses of different temperatures
collide with the warm air overrunning the cold. Turbulence
usually accompanies this condition and pollutants are scattered. The third and troublesome inversion type is the
subsidence inversion when a layer of air sinks, warms, and
remains stagnant over cooler air. (See Figure 8-3.) This
phenomenon can often be found in the northeastern United
States during the fall. In urban areas, the pollutants will
be trapped under the warm layer and may accumulate for days,
creating dark and noxious air. Respiratory patients will be
affected. If the inversion continues, it eventually becomes
a noteworthy episode.

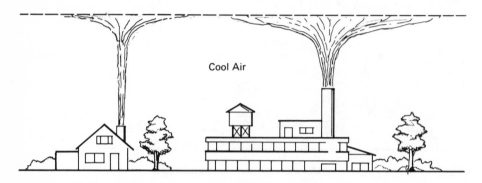

Figure 8-3. An inversion condition exists when a layer of warm air is above cooler air. Pollutants are then trapped in the lower zone. (From McCabe and Mines, Man and Environment, Vol. I, Prentice-Hall, Inc., 1973.)

Wind. We depend upon wind to disperse pollutants horizontally. If a city is advantageously "ventilated" with open space surrounding and frequent breezes blowing, as is the case of Boston or New York, pollution will rarely pose a critical crisis.

Wind will also blow pollutants from one population to another. It is possible to calculate the average wind conditions over a period of time for any particular site. And this will help in calculating the populations most affected by any particular population source.

Topography. Topography refers to the roughness or surface configuration of a terrain. Irregularities may be natural, such as hills, or man-made, such as buildings; but they will influence the air circulation. When an inversion occurs against a mountain or in a valley, air pollution will be intensified because the pollutants have neither vertical nor horizontal freedom of movement. (See Figure 8-4.) On

the other hand, a breeze against buildings can create turbulence enhancing the vertical mixing and dilution of pollution.

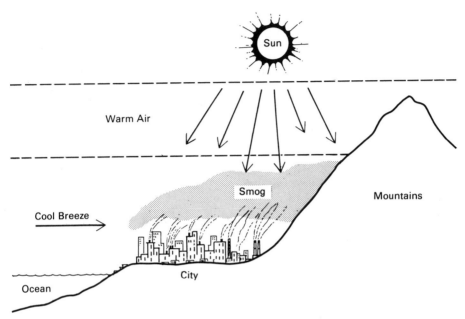

Figure 8-4. In the case of Los Angeles and areas of the
            Carolinas, inland mountains prevent the hori-
            zontal dispersal of pollutants. (From McCabe
            and Mines, Man and Environment, Vol. I,
            Prentice-Hall, Inc., 1973.)

Precipitation. Pollutants can be entrapped in precipitation either when the raindrop is being formed or as it
falls. Large particles are particularly effectively washed
out. Others will settle to the ground by gravity. Coales-
cence and chemical reactions will remove still others, and
some gas molecules will be removed by adsorption to particulates. Still other gases and particulates will react chemically to a form that will settle out of the air. In this
context, precipitation offers only one of numerous mechanisms that can potentially clear the air of pollutants. In
other words, the mechanisms for maintaining clean air are
available if the human factors, the propensities to pollute,
can be controlled.

QUESTIONS FOR DISCUSSION

1. Should urban residents receive a periodic warning that air-pollution conditions may be hazardous to their health? Explain.

2. Considering you to be an air-pollution expert, a city planning board has asked you to provide a list of three probable pollutants and associated health effects that could be expected from construction of (a) a new limited-access highway through the central business district and (b) a municipal incinerator in a mixed residential/commercial area.

3. Should an air pollutant be classified as "natural"?

4. Given your knowledge of surrounding communities, where would you expect to find exceptionally high levels of sulphur dioxide? Particulates? Carbon monoxide? Nitrogen oxide? Trace metals?

5. What are the arguments for and against forcing manufacturers of new chemical products to provide convincing evidence that the product's pollutants will not pose a long-term health threat in the exposed environment?

SUGGESTED READINGS

An exhaustive review of known information on the health effects of major air pollutants is contained in a series of criteria documents published by the government as authorization by the 1967 amendments to the Clean Air Act. A typical example would be:

1. U.S. Department of Health, Education and Welfare, Public Health, Air Quality Criteria for Carbon Monoxide (Washington, D.C.: U.S. Government Printing Office, 1970).

Numerous volumes present a more condensed description of the health effects.

2. George L. Waldbott, Health Effects of Environmental Pollutants (St. Louis: C. V. Mosby Co., 1973). 316 pages. An authoritative, readable and inexpensive example that systematically focuses on the sources, characteristics, and health effects of air pollutants.

Environmental
Quality
Management
Granville H. Sewell

3. Air Conservation Commission, AAAS, <u>Air Conservation</u>
    (Washington, D.C.: 1965), publication No. 80, 335
    pages. A helpful though slightly dated compendium
    prepared under the auspices of the American Association
    for the Advancement of Science.

    For an exceptionally concise, authoritative statement
of our current knowledge on the subject of inadvertent climatic change, refer to Helmut E. Landsberg's article.

4. Helmut E. Landsberg, "Man-Made Climatic Changes,"
    <u>Science</u>, Vol. 170, No. 3964 (December 18, 1970),
    pp. 1265-1275. Provides a detailed, interdisciplinary
    survey of the subject but is technical and less easily
    read by the nonscientist.

5. William H. Matthews, William W. Kellogg, and G. D.
    Robinson, <u>Man's Impact on the Climate</u> (Cambridge,
    Mass.: MIT Press, 1971).   594 pages.

Chapter 9

IMPROVING THE AIR QUALITY

> "I did inspect carefully . . .
> the gas works situated at the
> foot of East Fourteenth Street,
> in the City of New York, and
> . . . I have no plan to suggest
> by which we can be relieved by
> the present annoyance [from
> noxious gases], except the use
> of coals free from sulphur."
> Dr. Edward H. Janes (1866)

Scope of the Problem

Effective air pollution control requires orchestrating a multitude of talents. Deciding where, when, and how to measure pollutants in the air requires a sense of sociology combined with the technical skills found among chemists, statisticians, meteorologists, and, usually, several types of engineers. Identifying and measuring the pollutants that may exist in parts per billion of air challenges the skill of the analytical chemist. Appropriate control measures would probably include some prevention of the pollution, but this typically requires changing attitudes and habits of fellow citizens, a step that necessarily involves economic repercussions and political action. And the design of air-pollutant removal devices can require professional knowledge ranging from textiles through electronics. Because of variety, nature, and prevelance of sources, legal problems are potentially enormous.

These complexities probably explain why the control of air pollution has proven less tractable than many of the other environmental ills afflicting society. For example, numerous excellent textbooks on water pollution control exist, but no single, comprehensive textbook on air pollution has yet been widely recognized as authoritative. Thus, the field remains relatively unclaimed by any one professional group or approach. A few steps and general technical concepts, though, are universally accepted.

The Survey

Initial Information

As with any assignment to improve environmental quality, the first step is to investigate the problem. Two sets of information are needed: (1) quantities and sources of pollutants and (2) the objectives (standards) to be attained in air quality. A study of the sources and quantities of pollutants will provide the first clue to the types of control measures needed. And, when weighed in terms of total costs and benefits, the objectives--such as prevention of disease, protection of property, and preservation of the natural environment--will indicate the probable priorities that will be given to controlling specific pollutants.

Measuring Air Quality

The existing air quality situation is usually determined in two steps, (1) the emission survey and (2) air analysis and monitoring. Air analysis alone is insufficient. Pollutants may be emitted only intermittently, and meteorological conditions will cause amounts at any one point to vary widely. Furthermore, a monitoring program is easier to design if major pollutants are already known; and, for some relatively toxic materials, the trace amounts that are being measured are easier to analyze if special analytical preparations have been made.

An emission inventory is, in less sophisticated language, a survey of pollution sources and their pollutants. The method of collecting information will vary for each community. In most cases, old reports on air pollution will provide a beginning. City officials, especially the building inspectors and the building permit officers, can contribute clues. Fuel dealers and utility companies can identify levels and types of industrial activity. Traffic departments will have traffic counts over time. Citizen groups often have collected information on more obnoxious polluters. An interview with the major offenders, such as institutions and industries, may provide further insights. But from this information one should be able to prepare maps indicating the probable major pollutants and their levels for various times during the week, assuming stable weather conditions.

Sampling Plan

The usual purpose of air pollution control is to protect life and property from exposures, so estimates of exposures, not emissions, represent the ultimate step in defining an air pollution problem. In summary, exposure estimates can be made by taking air samples at points within a city or region, analyzing the chemical composition of the samples, and extrapolating from these analyses to obtain an estimate of exposure for something immersed continually in the air. For a sample to be valid, however, the three critical questions of what, where, and when must be answered.

What. The pollutant to be measured in an air sample partially determines the means used to collect the sample. So usually the emission survey plus some reasonable deductions based on experience and observations are used to select the pollutants for sampling. In a few cases, it may be desirable to "fish," to take whole air samples and analyze them for a broad variety of pollutants, first on a qualitative basis and then quantitative. But this approach is typically avoided because of the high laboratory costs.

Where. Since the air sampling occurs at one point or a series of points, the location for the sampling must be considered. In theory, measurement should be made at points of critical exposure. For people, this would mean measurements at nose level wherever they move, assuming that the pollutants are absorbed within the lungs. But sidewalks, offices, factories, and homes are not convenient areas for the bulky measuring instruments, and everyone breathes in different places, moving frequently from outdoors to indoors and from city to suburb. In practice, many sampling stations are located on roofs, or samples are taken through open windows. And, while measurements are taken in different areas of a city, estimates are often based on extreme cases, such as the person who lives and works within a zone of highest pollution concentration.

When. Sampling also occurs on a time dimension. If the probable time pattern of pollution is known, only a few samples taken during the day will suffice to confirm the pattern. Otherwise, sampling at brief intervals may be necessary to identify the times of lowest and highest pollution concentrations.

Types of Sampling Devices

Two fundamentally different types of sampling devices can be identified: particulate samplers and gas samplers.[*] By definition, particulates have a different chemical form-- liquid or solid--from the gaseous air media, and a wide range of particulate samplers have been developed. Considerable accuracy can be obtained, and particulate samplers can be found for almost any special purpose. In contrast, samplers for gases are relatively expensive, usually require a high degree of technical expertise, and the accuracy is often low. While numerous new gas sampling instruments are being introduced, the range for any one gas is still small compared to ones available for particulates.

Most devices for measuring particulates in the atmosphere can be classified within six categories.

(1) <u>Color Matching</u>. In the late nineteenth century, Professor Maximilian Ringelmann of Paris responded to the growing need for a means to quantitatively measure the black coal smoke that was annoying urban dwellers, so he devised the Ringelmann Smoke Chart that is still a standard device today despite the changes in the nature of air pollution. The Ringelmann Smoke Chart consisted of five (four now in the United States) rectangular darkness shades that can be matched to the darkness of the visible smoke.[**]

(2) <u>Settling Devices</u>. Since particulates are normally heavier than air, they will settle to the ground and can be collected by any object that may be placed in the open. All types of plates, jars, trays, and boxes are used. These are often coated with a jelly to prevent the particulates from being blown off. After a specified time, the particulates can be counted and removed for chemical analysis.

(3) <u>Impingers</u>. If air moves towards an object, it can change direction more easily than particulates, which will

---

[*]Most sampling devices combine the operations of removing the pollutant from the air and measuring it, but a few only collect the pollutant and another instrument is used for measurements.

[**]Shades are created by grids drawn with black lines of specified width and spacing or photographic film exposed to specific darknesses.

typically strike the object. If the object's surface is coated or prepared, the particulates will stick. They can later be counted and analyzed. But this system requires technical expertise and the smallest particulates are usually lost with the air.

(4) <u>Filtration</u>. Probably the most popular method of measuring particulates is to draw a known quantity of air through a filtering material. The material can then be weighed or the darkness can be measured automatically. If the quantity of pollutants is sufficient, the filtering material can be cut into strips or squares and analyzed for specific chemical constituents.

(5) <u>Cyclone or Centrifuge Devices</u>. If air is twirled in a circle, the heavier particulates will move to the outside and can be collected. Reliability is again low since many small particles are usually lost in the exiting air and interpretation of the results is difficult.

(6) <u>Light Beam Interruption</u>. If a beam of light is directed across a gap of known distance, the amount of light passing to the other side will depend upon the interference of passing particulates. A sampling device using this principle is easily adapted to automation, but since the light-scattering quality of particulates is dependent upon several factors, meaningful interpretations of the results are difficult to make.

By contrast, only three general techniques can be used to sample gases: absorption, adsorption, and direct physical measurements.

(1) <u>Absorption</u>. Absorption of the pollutant by another material that facilitates analysis is usually performed by one of two methods. Using a vacuum pump, the polluted air is drawn through cotton or other fiber material impregnated with a chemical that reacts with the pollutant to change color. The color is then compared to a master chart that shows the degree of color change associated with differing levels of pollution.

Alternatively, the polluted air is bubbled through water or a chemical solution where the pollutant can be chemically or electrically measured. The method is inexpensive and not complicated, but it is gradually being abandoned because of unreliability. Contaminating materials are also absorbed, and the degree of absorption is too frequently

uneven. In early 1972, for example, Environmental Protection Agency investigators discovered that the standard procedure for measuring nitrogen dioxide in the atmosphere may
have been indicating higher levels than actually existed.
Previously, they had assumed that the percentage of nitrogen
dioxide absorbed in a sodium hydroxide solution was the same
regardless of the concentration of nitrogen dioxide in the
air while, in fact, the percentage varied considerably.

(2) Adsorption. Adsorption, the adherence of a pollutant on the surface of another material, is used as both a
combined collection-measurement technique as well as simply
collection. Some pollutants, such as trace metals, can be
attracted to specially prepared surfaces, such as a gold or
silver foil, where they can be measured electronically or by
other means. More frequently, activated charcoal is used to
collect molecules of a pollutant; and the charcoal is later
heated or treated chemically to release the pollutants for
analysis. But neither technique avoids the problem of contamination, and a pollutant can be difficult to separate
from another material, especially charcoal.

(3) Direct Physical Measurement. Because of the many
difficulties encountered in the two-step process of collecting, then analyzing pollutants, techniques that use physical
or chemical characteristics of a pollutant to measure it
with almost no intervening steps are increasingly favored.
Chemiluminescence, the flash of light that some pollutants
release when they react with another gas, can be measured.
Nitrogen dioxide and ozone are particularly appropriate, but
the equipment is expensive and maintenance is constantly required. Various electrical and light-absorbing characteristics of other gases can be used. But no technique is yet
developed that combines low cost, reliability, sensitivity,
and ease of use. Lasers may eventually provide the key component for instruments that will have many of these characteristics.

Interpretation of Analyses

Air-quality measurements only provide clues to the
materials being inhaled by people as they move about a city
or region under myriad circumstances. Meteorological conditions when the samples were taken should be recorded so the
levels anticipated under the most adverse conditions can be
estimated. Time of sampling should also be recorded so cor-
relations with emission patterns and population movements

can be made. And exact vertical and horizontal location of sampling is vital information so extrapolation to other locations can be made and the sampling can be duplicated in the future.

In a specific study, a general understanding of the overall air-quality situation will gradually emerge. High levels of pollutants in particular areas will become associated with sets of sources. Some conditions will be deemed more serious than others. And the alternative steps that can be taken to abate pollution will begin to crystallize. These will exist at two levels, the technological solutions and the social or institutional strategies for motivating people to accept the technological solutions.

Alternative Technological Controls

Questioning Technology

The measures finally chosen for a pollution-control plan should be selected only after a thorough consideration of benefits and costs for all possible actions. Yet the range of possible actions to control air pollution is particularly sparse and uninviting, and there may be no other field of environmental management where as much attention has been given to either eliminating the need for a polluting technology or finding an improved substitute. Transportation provides a prime example. Tons of pollutants ranging from nitrogen dioxide to asbestos and rubber particles are being spewed into streets by increasing numbers of cars that are needed to span the increasing distances between residences and work places.

Expanding power plants are another common source of air pollution. Do we really need more energy? A similar rationale can be applied to the growth of factories, institutions, and residential units. New manufacturing processes can be developed, or new means can be found to perform functions that have previously required polluting technologies. With the fundamental cause for the pollution eliminated, the peripheral problems of management--including financial, law enforcement, and future adaptation--are eased, assuming that the new situation does not produce even more severe problems.

The "Technological Fix"

On a short-term basis, though, a device that can be attached to a pollution source or even placed in the air supply to reduce or remove pollutants will normally be sought. As in the case of sampling, there are two general classes of pollution-control technologies--particulate and gas. All tend to be technologically difficult, but this is particularly true for gases. The reason is simple. Polluting materials, such as sulphur, are usually present in small amounts within a fuel. When burned, the fuel combines with oxygen and expands manifold. The pollutant is now dispersed in an immense volume of air. Extracting it means the treatment of all that air. Just from the viewpoint of physical volume, the problem becomes staggering.

Particulate Control

In the case of combustion, particulates are usually controlled by one or more of the following devices: afterburners, settling chambers, cyclone separaters, baghouse or panel filters, scrubbers, and electrostatic precipitators. Each has a set of advantages and disadvantages, and each can be found used--or misused--in a particular situation.

Afterburners. Afterburners are normally associated with control of gaseous pollutants, but they may also be used for reducing particulates to a gas by completing combustion. In this case, the afterburner will be a hot chamber beyond the boiler or primary combustion chamber, and the particulates will be exposed to conditions--usually a catalyst or intense heat--that will cause them to burn further.

Settling Chamber. Gases usually leave a combustion chamber at high velocities because of the heat and expansion of fuels as they burn with air. If this gaseous mixture can be made to slow down, usually to velocities less than ten feet per second, particles will drop to the floor. So a settling chamber is usually a room or enlargement in the exhaust system from a furnace or other combustion area. Usually some means, such as trays or bins, are provided on the floor to facilitate removal of the ash or other particulates. This system is relatively inexpensive but will only remove the largest particles, such as those forty microns or larger in diameter.

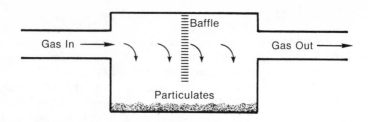

Gas In →   Baffle   Gas Out →

Particulates

Settling Chamber

<u>Cyclone Precipitators</u>. Cyclone
precipitators consist of a circular
space with the gaseous mixture being
sent spinning around the inside walls.
Being heavier, particulates will tend
to cling to the sides, slow down,
and eventually drop into a bin below.
Gases, however, can move to the center
of the space and rise upwards through an
exhaust system. Particles ranging from
five to several hundred microns in diameter can be removed. The device is simple
in concept, relatively inexpensive, and
not difficult to maintain or operate. But
its ability to remove the particulates can
vary tremendously, and, in the best of circumstances, the smallest and most dangerous
particulates from a health viewpoint remain
in the air. Sometimes the cyclone or the
settling chamber can be slightly improved in
efficiency by sonic collectors, high-frequency
sound waves that cause particles to vigorously
vibrate and collide, producing larger particles
that can be removed. But the sound energy
represents another hazard to operators, and
it is expensive as well as often lacking in
effectiveness.

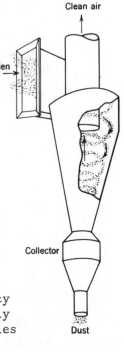

Clean air

Dust-laden
gas

Collector

Dust

Cyclone
Precipitator

<u>Baghouse Filter</u>. To a degree, the baghouse filter
resembles a huge vacuum cleaner. Gases are pumped through
the walls of tubular bags that catch the particulates.
Since the pores in the bags would eventually be clogged,
the gases are periodically diverted and the bags shaken so
the collected materials will fall into a bin and the filtering capacity restored. The efficiency of the baghouse
filter can be exceptionally high, but the filters can also

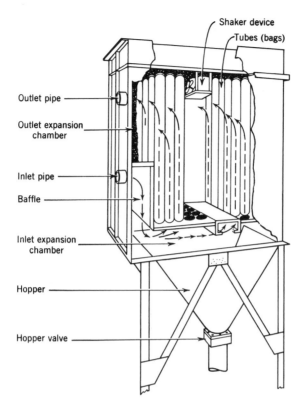

Shaker device
Tubes (bags)

Outlet pipe

Outlet expansion
chamber

Inlet pipe

Baffle

Inlet expansion
chamber

Hopper

Hopper valve

Baghouse Filter

be expensive to purchase and maintain. The hot air that is
sometimes laden with corrosive gases can quickly ruin the
fabrics. Nor does the vigorous shaking enhance their longevity. Furthermore, the pressures needed to force the air
through the bags require expensive pumping, but the collection efficiencies will vary according to the thickness of
materials clinging on the bag. Very small particles--those
below about five microns in diameter--can often pass through
the fabric alone.

In some cases, screen filters are used instead of bags.
The same problems found with baghouse filters usually apply
except that these screens cannot normally be shaken but must
be replaced frequently.

Scrubbers. If water or a chemical can be sprayed
through the gaseous stream, many of the particulates will
be washed out. But an air-pollution problem has merely been
transformed into a water-pollution problem. Furthermore,

the collected liquid is frequently
highly corrosive and can ruin the
pumping and spraying apparatus.
Since the gaseous stream is now
cooled and moisture-laden, it may
settle quickly to the ground and
cause a concentrated pollution
problem locally. Collection efficiency varies widely depending
upon the elaborateness of the
system; but under some circumstances the efficiency can approach 98 or 99% removal.

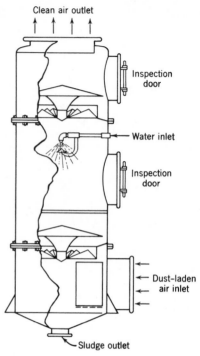

Scrubber

Electrostatic Precipitator.
Thales of Miletus in the fifth
century B.C. is credited with
first recording the phenomenon
that a charged material (amber)
will attract small particles.
This principle that opposite
electrical charges attract is
utilized in the electrostatic
precipitator by passing the
particulate-laden air between
charging plates with voltages
of 25,000 volts or higher and
then past collection plates or
electrodes with the opposite charge.
Receiving an initial charge at the first stage, the particulates are attracted to the collection plates, are stopped,
and then fall into bins where they can be removed. Because
electrostatic precipitators are effective with small particulates, they are usually preceded by a cyclone precipitator,
settling chamber, or similar device that removes larger particles.

The electrostatic precipitator is capable of exceptionally high collection efficiencies, 99% or better. But it
has numerous weaknesses. Because of the high voltages
needed, equipment is expensive to install. Because the
plates and electrodes are exposed to the airstream, they
need constant maintenance and frequent replacement. Poorly
adjusted precipitators can create an ozone problem. And
even an efficiency of 99% for a 1,000-megawatt generating
plant burning coal can still mean over six tons of

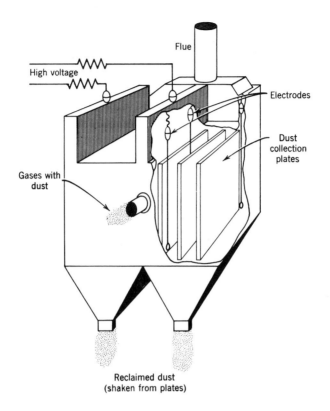

High voltage

Flue

Electrodes

Dust
collection
plates

Gases with
dust

Reclaimed dust
(shaken from plates)

Electrostatic Precipitator

particulates being released to the environment daily, and
these will be the smaller, more dangerous sizes.*

In practice, studies have discovered that few electrostatic precipitators actually operate at their design
levels. Inadequate design and sizes or lack of maintenance
appear to be the usual causes for low efficiency. Operating
at full capacity for twenty-four hours, the hypothetical
plant described above could raise its particulate release
to over 120 tons daily assuming 80% efficiency.**

---

*Estimate assumes about 10% ash content for the coal and a
total production of about 67,000 pounds of particulates per
hour.

**Charles Komanoff, et al, The Price of Power: Electric
Utilities and the Environment (New York: The Council on
Economic Priorities, 1972).

Gas Control

The alternatives for controlling gaseous pollutants after combustion are few and generally unsatisfactory. Afterburners can sometimes help if the polluting gas can be further burned, but the usual techniques involve either absorption or adsorption.

Absorption. In an absorption process, the gas is passed over water or a chemical that will remove the pollutant. Since all molecules of the gas should come in contact with the liquid, a tower is usually used and the gas is pumped through sprays, screens, or some type of packing material. The process is slow, energy consuming, and normally expensive.

Adsorption. Some gases can be collected by exposing the molecules to an adsorbant, including activated carbon, silicates, aluminas, or special gels. The gas can later be removed with heat or steam, and the adsorbant can be reused. But one still has the gas molecules to treat in some manner, and the process usually has low collection efficiency and high expense.

Packed Tower

Implementation

Determining Strategy

From a study of each alternative, a pollution-control plan will gradually emerge. A few steps will be relatively easily taken. Political pressures and publicity can be focused upon public agencies and institutions--such as municipal incinerators, hospitals, and public housing developments--that are significant polluters. Personal persuasion may sway some industries, but the long-term effort will invariably have to concentrate upon public education and regulation. On a national level, several decades of experience have been gained in fashioning and applying legal instruments to control air pollution.

Legal Instruments

Background

Air-quality legislation represents the control of social behavior to prevent or reduce the effects of air pollutants on the environment. Historical precedents date from the Middle Ages, and an extensive legal tradition has been developed. But the nature of the air-pollution problem shifted drastically after World War II with the burgeoning use of automobiles and the replacement of coal by petroleum in space heating and industries. So the society has been forced to develop a new set of legal instruments, including the public support and the governmental structure needed for implementation. In essence, the tendency has been to move steadily from the smaller political unit to the larger, from the local to the regional, and, eventually, to the national. And the focus has gradually shifted the problem of modifying or eliminating specific technological devices.

The legal structuring in the United States began when complaints by merchants and housewives over the soot and grime being spread by coal burning, especially from steam locomotives, caused numerous cities during the late 19th Century to pass smoke-control ordinances. By the mid-20s, more than fifty cities had these regulations, and an enforcement division was established under the jurisdiction of health, building, or public safety departments. The problem began to subside as oil replaced coal and automobiles supplanted trains. Then, as the traditional pollutants disappeared, new ones resulting from high-temperature combustion of petroleum appeared.

The automobile·spawned the spread city that spilled across traditional city boundaries, so the legislative initiative passed to the state. Prodded by the special problem in Los Angeles, California passed the first state air pollution law in 1947. Oregon soon followed. The Donora episode heightened public concern, but the movement remained slow as states persisted in viewing the problem as local. By 1963, only fourteen states had laws that considered air-pollution control on a statewide basis, although two others had limited coverage laws. Meanwhile, the federal government was trying to stimulate the states into action.

Between 1950 and 1954, many unsuccessful resolutions and bills were introduced into Congress. One entitled "Air Pollution Control--Research and Technical Assistance" was finally enacted in 1955 to provide state and local air-pollution control agencies with research, training, and demonstration assistance. It also authorized a small federal role in collecting and publishing air-pollution information and providing grants for technical advice and assistance. But the law explicitly stated that the prevention of air pollution was a responsibility of state and local governments. Even the five million dollars each year authorized by the bill was unspent since President Eisenhower and his budget advisors opposed federal involvement.

With each passing year, the need for further federal legislation to counter the new pollution was becoming more obvious. President Kennedy pressed for action, but the bill that he urged was not passed until 1963 after President Johnson took office. While still stating the principle of state sovereignty, the Clean Air Act of 1963 (P.L. 88-206) provided for the first time that the federal government could take direct action to correct an air-pollution problem. The act was almost ridiculously weak, though. The Public Health Service could act without state permission only if population in more than one state was affected and then the procedure was limited to hearings, conferences, and requests for court action. But the need for regional planning and control in abating air pollution was recognized; the principle of direct federal involvement was established; and more generous aid to states was provided.

By 1965, the consumer movement was beginning to publicize the role of the automobile in worsening air conditions; and Congress responded by passing the 1965 amendment, sometimes known as the Motor Vehicle Air Pollution Control Act (P.L. 89-272), to the Clean Air Act. The Secretary of Health, Education and Welfare was given authorization to establish permissible emission standards for new motor vehicles, a step that the automobile industry supported because it prevented states from taking individual action. (California was the sole exception.) As another compromise measure, the Surgeon General was directed to study the health effect of automobile emissions.

A logical next step would have been the establishment of national emission standards for major industries and a degree of federal enforcement. But industry, particularly the coal interests, which felt threatened by possible stricter sulphur dioxide standards, exerted immense Congressional pressure. The House of Representatives is normally more conservative than the Senate on environmental matters, and Senator Jennings Randolph of West Virginia was Chairman of the Public Works Committee. Since he had higher jurisdiction over Senator Muskie's Committee, which was preparing the new bill, a compromise was inevitable. So the Air Quality Act of 1967 (P.L. 90-148) authorized the Department of Health, Education and Welfare to define air-quality control regions, issue air-quality criteria, and provide reports on control technologies, but the state governments--after hearings that followed a specified procedure--would set the air-quality standards and be responsible for enforcement, although the federal government had powers of review.

The Air Quality Act did not have the direct powers to produce drastic improvements in air quality, but it did contribute indirectly to the passage of stricter legislation. One unexpected factor was the required setting of standards at regional hearings that raised the anger of many citizens by exposure to abrasive industry representatives. Questionable industry tactics convinced a large sector of the public that reasonable air-quality goals could not be achieved with the attitudes existing in some industries and the federal powers of enforcement.

A case example was the struggle to stop one small Maryland chicken renderer from sending a sickening stench over the Delaware border to the town of Selbyville. The case, now known as the U.S. vs. Bishop Processing Company, began in 1956, when the company received an order to stop emissions. Over the next nine years, both the state of Maryland and the state of Delaware attempted to stop the offensive aspects of the plant operations. The case then became the only one to go to court under the 1963 Clean Air Act. The federal government called a conference, and a consent decree was obtained in federal court in 1968. An order was issued in 1969 to stop operations. The case was appealed and it went to the Supreme Court in 1970. The plant was then ordered unequivocally to cease all operations forever. But, according to a Washington Post article of July 29, 1971, the plant was still in operation. The company was then held in contempt of court, but no penalty was ordered. Presumably, the plant is still operating.

With such examples, Congress reconsidered the air quality legislation and passed the Clean Air Act Amendments of 1970 (P.L. 91-604). The entire country was divided into about 250 regions based on climate, meteorology, topography, urbanization, and politics. The administrator of the Environmental Protection Agency was ordered to establish national primary and secondary air-quality standards. Primary air-quality standards were those that, in the judgment of the administrator, were necessary to protect the public health while allowing an adequate margin of safety. Secondary standards were those that would protect the public welfare, which was defined as being property and the natural environment.

In addition, the states were required to develop implementation plans, and the EPA Administrator was given strong powers of control. Primary standards were to be achieved in three years after acceptance of the plan, and the date that emerged was about 1975. The automobile was especially cited, but the administrator could also insist upon the best available system of emission reduction being installed for all new stationary sources of air pollution, such as industries and power plants.

The amendments seemed to be everything that environmental advocates sought. But the victory soon began appearing hollow since it was apparent that enforcement in all aspects was impossible. To meet the national standards would require widespread abandonment of automobiles in dense urban areas where rapid transit alternatives had not been developed. Also, the new pressure on energy resources made the standards difficult to defend when the alternative could be portrayed as cold homes and public institutions. In another sense, however, the environmental advocates did achieve an important gain because the general public began to realize that some individual sacrifices and adaptation would be necessary if the generally acclaimed environmental goals were to be achieved. Over a long term, creation of this awareness was an absolute necessity.

QUESTIONS FOR DISCUSSION

1. Outline the initial research you would perform before
    formulating an air-pollution abatement program for your
    residential community.

2. Assuming you want to reduce pollution from a municipal
    incinerator, identify three air-pollution control de-
    vices you could use and the pollutants you would expect
    to remain in the air stream after this treatment.

3. What opposing and supporting forces--social, political,
    and technical--would you have to consider in establish-
    ing an air-pollution control program in your community?

4. Should all air pollutants be considered "guilty" and
    subject to control regardless of the lack of information
    available describing anticipated effects? Consider the
    political and economic implications.

SUGGESTED READINGS

    The standard reference for technical problems in air-
pollution control is still:

1. Arthur C. Stern (ed.), Air Pollution, 3 volumes (New
    York: Academic Press, 1968).

    The government publishes some engineering manuals, and
some technical volumes are being marketed in competition to
Arthur C. Stern's Air Pollution.

2. Jack A. Danielson (ed.), Air Pollution Engineering Man-
    ual (Cincinnati: Public Health Service, H.E.W., 1967).

3. Werner Strauss (ed.), Air Pollution Control, 2 volumes
    (New York: Wiley-Interscience, 1971).

    The American Public Health Association publishes a
thin, dated but helpful booklet on the planning aspects of
air-pollution control.

4. APHA, Guide to the Appraisal and Control of Air Pollu-
    tion (Washington, D.C.: APHA, 1969).

Chapter 10

NOISE CONTROL

> "There above noise and danger
> Sweet peace sits crown'd with
> smiles." Henry Vaughan (1655)

Basic Concepts

Physical and Psychological Dimensions

Noise is commonly defined as "unwanted sound." This definition embodies the two inseparable elements of noise-- judgmental preference (unwanted) and physical phenomenon (sound)--that makes its control both necessary and frustrating. Physically, noise is only one type of sound. Sound, the oscillation of air pressure causing an auditory sensation, can be measured and analyzed interminably without anything distinguishing it as noise. The key difference between sound and noise is the "unwanted" aspect. An individual human judgment must make the distinction. One person's noise may be another's music.

Dual Analysis

At acute levels of noise, disagreement over its offensiveness tends to fade--but not quickly. For example, noise that is sufficiently loud to hurt or cause deafness would presumably be avoided or prohibited. But discotheques often have noise levels that would cause a factory to take immediate protective measures or be closed as hazardous for the health of workers. So noise has to be identified by two analyses, one of sound and the other of offensiveness or, as it is commonly called, annoyance.

Physical Characteristics of Sound

Pressure Waves

Sound is transmitted by pressure waves through air travelling from some vibrating object to the ear of a

listener. Pressure waves can be produced in any material--
water, gelatin, even metals--that has mass and elasticity.
Air has mass in the same sense as water or metals. A typical living room will contain about 150 pounds of air. Air
also has considerable elasticity, as anyone who has pumped
a tire with a hand pump can testify.

To describe what happens when sound is produced, we can
consider the movement of a vibrating object, such as a harp
string. When moving against the air in the direction of the
listener, the string will compress some air molecules, which
will--in turn--push against other molecules and thus cause
the pressure wave to move through the air like the moving
collapse of a row of dominoes. As the compression wave
passes the ear and activates the auditory system, sound is
registered. But the string does not continue to move in a
single direction. In a fraction of a second, the string is
moving in the opposite direction and compression towards the
listener changes. Then in another fraction of a second, the
object is again compressing air towards the listener and
another compression wave is moving towards his ear. Thus,
the phenomenon that we call sound consists of repeated pressure waves striking the ear.

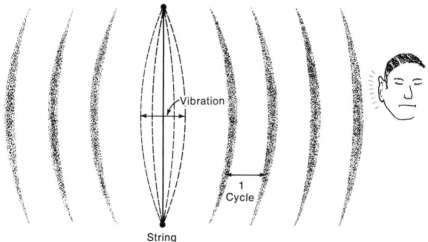

Figure 10-1. A vibrating string causes a series of pressure waves to move towards the listener.

Two characteristics--loudness and pitch--of sound are
primarily used to describe it. Loudness is caused by the
pressure or push of the pressure wave against the ear or
other measuring instrument. The stronger the pressure,
which is dependent on the speed and distance of the vibrating

object's push against the air molecules, the louder the
sound. Pitch is determined by the frequency of vibrations,
the number of times the vibrating object pushes against the
air towards the listener in a second or any other period of
time. The more frequent the vibrations, the shriller the
pitch. All measurements of sound, therefore, must collect
information about either pressure (loudness) or frequency
(pitch) or, more commonly, both.

Pressure Measurements

When we try to measure the sound pressures that are
normally heard by the ear, two methodological problems arise:
(1) the immense range of audible sound pressures and (2) the
relative insensitivity of the ear towards pressure changes
when pressures are high. Both problems must be considered
in devising a system for measuring loudness of sound.

The normal ear can detect sound pressures as light as
0.0002 microbars to as high as 100,000 microbars before the
ability to hear is destroyed. A microbar is defined as the
pressure of one dyne over a square centimeter. A microbar
is an extremely small unit, somewhat comparable to the pressure exerted by a mosquito sitting on your arm.[*] But any
scale that uses a billion units to measure a common phenomenon is unwieldy.

The human ear solves the difficulty by being more sensitive towards lighter pressures (softer sounds) and less
sensitive towards higher pressures (louder sounds). In
mathematics, this is known as a logarithmic function. Like
a slide rule, the units are large and easily distinguished
when numbers are small but units become smaller and less
easily distinguished as the number rises.

To mimic the ear, we have arbitrarily defined a scale
for measuring noise known as the decibel, which is abbre-
viated as dB. The equation for translating sound pressure
levels into decibels is:

$$\text{SOUND PRESSURE LEVEL} = 20 \log_{10} \frac{P}{P_0} \text{ dB.}$$

P is the pressure measured by an instrument.

---

[*]The official definition of a dyne is the energy required to
move one gram a distance of one centimeter.

P_0 is a reference pressure normally defined as the lowest audible pressure, or 0.0002 microbars.

In a typical urban environment, the ear will be registering sounds from about 25 dB to 100 dB. Sounds in a quiet bedroom will not exceed about 35 dB, normal conversation will reach about 65 dB a yard away, and a large truck accelerating may produce more than 100 dB as measured from the curb. Even sound levels from a normal automobile may attain 90 dB during acceleration. Many small home appliances, such as a vacuum cleaner, or dishwasher, will exceed 70 dB when measured several yards away.

This scale has both convenient and inconvenient aspects. It does measure audible sound with the lowest feasible sound approaching 0 dB. Even the sound of a Saturn rocket lifting from a launching pad 100 yards away would not exceed about 200 dB. Thus, the length of scale is reasonable. The difference between measurements of pressure and dB, however, is confusing for the general public. Doubling the pressure increases the decibel levels by only approximately 6 dB. And a quadrupling of pressure would raise the decibel levels by only about 12 dB. In terms of hearing, though, about 10 dB increase is necessary to make a sound seem twice as loud to a listener.

Measuring Pitch

Pitch, the shrillness of sound, is measured in cycles (back-and-forth movements in air pressure) per second (cps) or, more officially, hertz, which means the same thing. The ear is more sensitive to rapid oscillations in air pressure than slow. In a physical sense, this is logical because rapid oscillations transmit more energy than fewer oscillations at the same pressure. In order to determine the effects of pitch on apparent loudness, people were tested in a laboratory to define loudness-pitch combinations of equal apparent loudness. If people were given a low-frequency sound and asked to adjust volume (loudness) so apparent loudness remained the same as the pitch steadily increased, an equal loudness level curve or phon curve was defined. This has been performed extensively, and a standard set of equal loudness curves has been published.

To adjust the decibel for pitch, a scale known as the dBA has been defined, and most noise ordinances express noise standards in terms of dBA. Sound-level meters generally have a setting marked "dBA" or "A Scale," and this

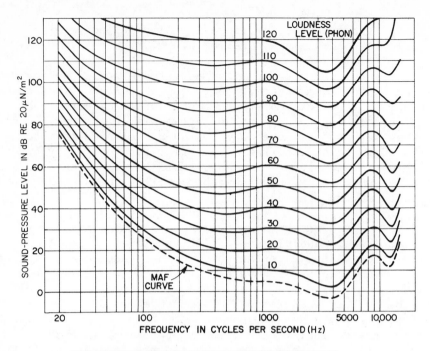

Figure 10-2. Equal loudness curves for pure tones as determined by laboratory tests. (From Handbook of Noise Measurement, 7th edition, 1972, General Radio Company. Contours
determined by Robinson and Dadson at the National Physical
Laboratory, Teddington, England, ISO/R225-1961.)

automatically registers shrill noises as having a higher
apparent loudness. To provide another scale that will be
almost insensitive to pitch, instruments use the dBC scale,
which is almost plain dB. There is often also a dBB scale,
which is a compromise between dBA and dBC, but this is rarely used. Thus, a shrill noise being measured will register
significantly higher on the dBA scale than on the dBC and
slightly higher than on the dBB.

Measuring Annoyance

Researchers soon realized that loudness was not synonymous with annoyance of a particular sound. So people were
tested in the laboratory for sounds of equal annoyance in a
manner similar to the one used for equal loudness. The

206

resulting scale is used for calculating perceived noise
levels measured in PNdB with units known as noys. Because
pure tones are rare, a method for averaging mixed noises
had to be devised and the result was a scale known as the
effective perceived noise levels measured in EPNdB. Numerous other scales have been constructed to account for duration of sound, mixed sounds, and other aspects; and still
other scales may be devised in the future. But the basic
scales will typically be those described in this section.

One type of sound commonly encountered but difficult
to measure is an impulse sound, sound produced by a single
primary vibration, such as occurs when two hard objects
collide. Because of the slow reaction time of most sound-
measuring equipment, typical measurements may underestimate
the sound level by 20 or 30 dB. Special instruments are
available to overcome this problem.

Objections to Community Noise

The Need to Register Objections

Everyone is immersed in noise, but few people try to do
anything about it. People hesitate to complain about community noise. City councils avoid noise ordinances because
their public appeal does not justify mastering the complexities of the subject. Even courts have tended to levy ridiculously small fines--five dollars has been a common fig-
ure--for violations of noise ordinances because of the legal
viewpoint that damage should be measurable in financial
terms. So a careful cataloguing of objections to community
noise is usually necessary if a noise abatement program is
to be successful.

Hearing Loss

Loss of hearing has been the traditional justification
for controlling noise. The term "boiler maker's ear" suggests some of the folklore behind this valid but complex
objection. Loss of hearing can have many causes, and at
least three categories are usually recognized. A conductive
loss means that the sound pressures are not reaching the
nerve centers in the inner ear. The most frequent cause is
the accumulation of ear wax, but the mechanical system for
transmitting the pressures from the ear drum to the nerve
centers may also be at fault. This type of hearing loss is

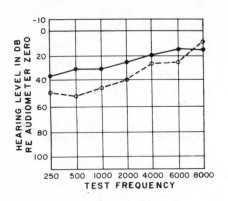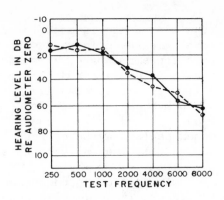

Figure 10-3. Left--Audiogram (hearing test) showing conductive type of hearing impairment. Note that hearing level for each ear (each line representing hearing on one side) rises slightly for increases in pitch, which is a normal tendency. However, the ears are relatively insensitive to all sounds. Compare the hearing at lower frequencies with that on the right.
Right--Audiogram showing a perceptive type of hearing impairment. Note drop in hearing for higher pitches.

often correctable, and it generally affects all frequencies of sound.

A second category is a perceptive (neural) loss, which means that the nerve cells are no longer capable of transmitting the pressure sensation from the inner ear to the brain. This is not reversible. The third category is functional loss, meaning that an individual has no physical impairment but, for psychological reasons, he cannot perceive the signals to the brain. In common language, he is tuning out the sound. Of these, excessive noise only affects the second category, the perceptive or neural loss, by causing nerve cells to atrophy, possibly as a defense mechanism against the excessive energy being transmitted to the brain center. Usually this loss is predominately in the upper frequencies of the hearing range.

Although we can identify the type of hearing loss characteristic of noise damage, at least three complications characteristically arise when we try to incriminate a particular noise source as the cause. (1) Hearing loss normally occurs with increasing age. Thus, it is difficult to prove that a particular loss was due to excessive noise and not to a natural aging process. (2) Hearing loss after exposure to loud noises can be temporary. Recovery begins

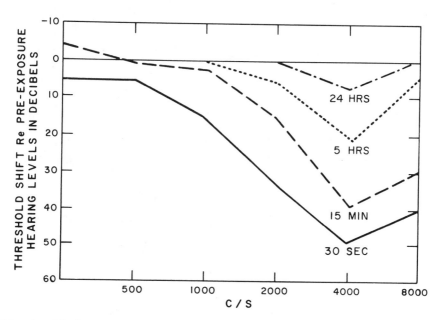

Figure 10-4. Changes in hearing thresholds relative to
pre-exposure values for different frequencies
measured at various times after a twenty-
minute exposure to 115 dB broadband noise.
(Alexander Cohen, Industrial Noise, p. N-4-1.)

almost immediately, and, if the exposure is not one that
occurs frequently or is excessively high, recovery will
almost be complete within one or two days. The last part
of the hearing ability to return will be in the higher
frequencies. (3) If exposures continue, the recovery will
become slower in the higher frequencies and, eventually, the
hearing loss will become permanent. Hearing loss is usually
selective. This leads to some deceptive phenomena. A worker with hearing loss in the upper frequencies may be able to
conduct a conversation in a noisy environment more easily
than a person with normal hearing because his ability to
hear the lower frequencies of a conversation are not impaired by extraneous higher-frequency noises. Nevertheless,
studies have demonstrated that rural residents have keener
hearing in all frequencies than urban dwellers, suggesting
that excessive urban noise does impair hearing and our ability to enjoy one dimension of a quality life.

Functional interference of noise with normal activities
is often used as a basis for setting noise standards. When
noise interferes with normal conversation, some loss can
be inferred. This becomes even more dangerous if the

nterference is with commands or warnings about impending
dangers. Economic losses can be shown if the interference
is with instruction or orders in a work place.

Noise will also lower working efficiency. Numerous
tests have shown that accident rates and, usually, productivity can be improved by lowering noise levels. It is very
difficult, though, to separate morale factors from effects
of noise. For example, a group of typists improved typing
speed when noise levels were deliberately reduced, but the
higher speeds were maintained even though the original noise
levels were gradually restored. In another case, similar
results were obtained with operators of film perforating
machines. Yet, in this case, the rate of machine stoppage
was shown to be directly dependent upon noise levels.

Physiological Effects

Physiological effects remain among the least understood
aspects of noise impact upon a community. Methodologically,
noise should be separated into two categories, meaningful
and meaningless. Meaningful noise, such as a baby crying or
a person screaming, acts as a stimulus on the body, and the
pattern of stress reactions is similar to that caused by
other stimuli.

Meaningless noises, though, are usually more frequent,
and the effects are more distinctive, although interpretation is still being debated. When exposed to an intense
meaningless noise, the peripheral circulatory system under-
goes vasoconstriction. Blood volume in the skin is reduced,
stroke volume of the heart decreases, and the pupil of the
eye dilates. To some degree, effects can be cumulative.
Sleep can also be affected by noise even though the person
may not awaken. Attempts have been made to associate physical illnesses, such as ulcers, with meaningless noise exposures, but the studies have not been particularly conclusive.
Yet the general assumption is that continual stressing of
the physical system is harmful and will contribute to low-
ered resistance towards other ailments.

Psychological Effects

Noise is inseparable from the rest of the environment
in producing psychological states. In general, though,
noise--by definition--is considered inherently annoying. At
least four conditions contribute to the degree of annoyance.

(1) Noises with unpleasant associations annoy more than neutral associations or meaningless noise. For example, the sound of an ambulance siren or a low-flying aircraft will produce anxiety and relatively high annoyance levels in most persons.

(2) A noise that is inappropriate for the activity at hand will be more annoying than one that is expected. This is particularly true during sleeping hours. A person sleeping generally does not expect to hear delivery trucks arriving or rubbish collection occurring.

(3) Noises that are considered unnecessary or contain no advantage can rate high in an annoyance scale. A person accepts his dog's barks but will object to the neighbor's dog. Few people will complain about the noise of an air conditioner, although many are noisier than necessary.

(4) Noises affect each individual differently. Some individuals will have an extreme anxiety over noises they interpret as threatening, and noise may become an obsession that affects their physical health. Others may simply find most noises to be annoying stimuli. Sleep will be interrupted, and attention to work problems may be diverted. Yet other individuals exposed to the same noises may not have any perceptible reaction. The sound may not enter their consciousness; and, if it does, it does not seem a significant stress factor.

Control of Noise

The Surveys

Like the definition of noise, the survey to define a noise problem should have two dimensions, the sound survey and a community annoyance survey. The objective of the sound survey is to identify the primary noise sources. Since noise measurements are relatively simple and inexpensive, they can be made on a considerably more systematic basis than for air quality. Different areas of a city can be surveyed for different times during a day. Special attention can be given to known noise sources, including subways, street traffic, aircraft, and construction equipment. Besides providing a factual basis for future decisions, the sound survey can serve as a training opportunity for personnel who may later be involved in the enforcement stage.

Annoyance surveys consist of interviews of community residents to determine the noises they consider annoying and, as much as possible, to establish degrees of annoyance. In some cases, subjects may be invited to a local noise laboratory where responses to possible changes in community noises can be tested. These annoyance surveys are not simple to conduct or interpret. Some people always resent the intrusion of an interviewer, and there is often a tendency to provide answers that are expected to please the interviewer. Furthermore, attitudes at any point in time can be easily influenced by recent events, and the persons being interviewed may not be conscious of all the troublesome noise sources. Because of these problems, some authorities tend to discount annoyance surveys, but a suitable substitute has never been offered.

Noise Minimization

In defining technological alternatives for decreasing noise levels, eliminating the annoying technology is an obvious but rarely-used technique. Noise reduction is too simple and inexpensive. Owners of a noisy machine prefer making it quieter to junking it.

In some cases, though, the replacement of a nearly obsolete technology may be hastened. Excessive noise has contributed to the dismantling of elevated trains in many cities. Small electrical generating units have often drawn effective criticisms from neighbors. But a more likely achievement is the isolation of a noisy technology. Traffic is often detoured, sometimes on circumferential highways around urban centers. An advantage frequently cited for new subways is the reduction of community noise.

Isolation can occur on a time basis as well as geographic. Traffic is sometimes only diverted at night, and night deliveries are often discouraged because of noise factors. Construction equipment at night is sometimes prohibited, and early morning waste collections are often rescheduled.

Modification

Some reduction of noise can often be achieved by ridiculously simple modifications of the noise-making device. Vibrating panels can be secured to frames or stiffened with ribs and matting. Truck tires can be redesigned. The shape of exhaust baffles and openings can be changed.

Each additional reduction in noise levels, however, becomes increasingly expensive. For example, mufflers for internal combustion operate by limiting some of the gas flow. This produces back pressures in the engine and efficiency drops with the noise levels. So the difficult task is to balance the amount of noise reduction desirable in a particular situation with the costs of that reduction.

Interception

Instead of trying to modify the noise-making equipment, some type of shielding can be placed between the machine and the listener. From a moral viewpoint, covering the machine with shielding, such as a building, would represent the least intrusion upon the listening bystander.

In other cases, though, some type of shielding in the area between the device and the listener will be necessary. Rows of trees along a highway can serve this function. Buildings are often constructed along the perimeter of a university campus to act as a shield against noise and other distractions. If a highway must pass through a crowded community, sloping berms (shoulders) for a slightly depressed highway will tend to reflect noise upwards from both motorists and house residents.

Enclosure of Listener

As a last resort, the listener can be enclosed. If he is primarily inside a building, the passage of noise through windows and walls can be reduced. As a general rule of thumb, noises can be reduced by as much as 10 dB by simply closing windows. Storm windows provide further protection. Thicker concrete or masonry walls can be used, and this practice is considerably more established in Europe than the United States. For interior partitions, heavier materials or the use of a construction technique that does not allow walls to vibrate as a single unit will contribute to lower interior noises.

If the listener is outside or enclosure does not succeed in reducing noise sufficiently, some types of ear muffs or plugs may be needed. As a rule, ear muffs are preferable to plugs. To be comfortable, ear plugs should be designed on a custom basis for each individual. The shape of ear channels vary between individuals and groups, such as men and women. Also, hygiene for ear plugs is important since the use of dirty plugs or their insertion into a wax-filled channel can cause an infection. But, outside of an occupational situation, use of personal devices for noise protection should only be moves of desperation.

Masking of sounds is a technique that may also be used under some circumstances. Background music, for example, disguises operational noises in a grocery or some factories.

Strategy

Legal Instruments

Until the mid-sixties, noise control was considered almost exclusively a local community problem, but local communities were not prepared or able to cope with the problem on a systematic basis. Regulation of aircraft noises, for example, was largely preempted by the federal government. Encouraging manufacture of quieter trucks and cars would have been a futile gesture by single towns or cities since their market share would have been too small to influence industrial decisions. Even when laws were passed, they tended to be weak. Judges often dismissed cases because of vague wording in laws referring only to unnecessary or unreasonable noises. Quantitative standards were lacking. Even when enforced, the penalties—usually a fine of five or ten dollars—were scarcely a deterrent for any industry.

The enforcement problem had deeper roots. Local authorities did not understand sound measurements or their relationship to annoyance. Sound-measuring equipment was poor in quality and expensive, so few communities would make the investment even if technical personnel were available.

Although most of these deficiencies at the local level remain today, the situation has dramatically changed with the entrance of state and federal authority. New York became the first state to enact a state highway noise-control law in 1965, although it has not been vigorously enforced. California followed with a more stringent statute in 1967 providing standards and test procedures for virtually all vehicles. By the early seventies, almost every state had some type of noise-control legislation, and many had passed laws describing the authority that could be assumed by local communities.

Though directives were issued in several federal departments earlier, the major federal entrance into the community noise problem was the Noise Pollution and Abatement Act of 1970 which was part of the Clean Air Amendments. An Office of Noise Abatement was established in the Environmental Protection Agency, and a report to Congress on noise problems and impacts was requested. This report was released in early 1972 after public hearings in eight major cities, and Congress subsequently enacted the Noise Control Act of 1972.

Through a complex system of action and reaction, the EPA and the FAA (Federal Aviation Administration) were required to issue regulations for aircraft. For railroads and interstate trucks, the EPA was to establish noise emission limits with enforcement being performed by the Department of Transportation. For all other major noise sources--construction equipment, home appliances, and transportation equipment--the Administrator of EPA was required to promulgate initial noise limits by October, 1974, if standards were feasible. He also was given discretionary authority to regulate other products where public health or welfare may be endangered. Most observers of this legislation have doubts about its effectiveness, however, because basic knowledge of noise effects on public health and welfare is still rudimentary.

Local initiative for most noise-control measures remains, though, since the federal law is restricted to regulation of new products. Gradually a pattern for community noise legislation is emerging. Quantitative standards are provided, usually about 75 dBA at fifty feet for daytime traffic to 70 dBA at night. Standards are about 20 dBA less if measured indoors. Unnecessary sounds are prohibited. Explicit authority for enforcement of the laws is lodged in one agency, such as the police department or health department. Some noise sources, such as construction equipment, may be prohibited from operating during night hours. And these measures may be supplemented by building codes that stipulate construction techniques that minimize noise transmission to building interiors.

Occupational Noise Exposures

Hearing loss from excessive noise has traditionally been regarded as a common but preventable occupational hazard for many industries. Special legislation has repeatedly been directed towards this problem, each effort being more stringent. The current federal legislation is the Occupational Safety and Health Act of 1970, which specifies the mechanisms for establishing safety and health standards for major employers. Interim standards have been set at 90 dBA for the maximum sound levels throughout a place of work unless personal protective devices (ear plugs or ear muffs) are made available and used. Other standards for intermittent noise exposures have been issued, and lobbying pressures for lowering the primary standard to 85 dBA have been intense.

Where excessive noises exist, employees are supposed to receive regular hearing tests, records are to be maintained, and a program of noise abatement is to be undertaken. Unfortunately, enforcement of the law has been less than vigorous, and progress is generally admitted to be slow.

Conclusion

Sound can be a deceptive stimuli. It is invisible and involves trivial physical forces. But some sounds can soothe and other sounds can arouse. Gentle sounds are associated with peace, warmth, and tranquility, while harsh sounds raise anxiety and anger. Physically, the full implications of these phenomena are still unknown. But in a hectic world paced by technology, preservation of some quiet becomes precious, and judicious control over noise generation becomes a necessary ingredient for sanity.

QUESTIONS FOR DISCUSSION

1. Using sounds from your community, give examples for each of the following:

   A noise conveying an unpleasant association.

A noise inappropriate for the activity at hand.

A noise considered inessential.

2. If a 100 hertz sound wave is travelling 1,128 ft. per second, what is the length of the sound pressure wave (crest to crest) in feet?

3. Why do the technological problems of noise measurement make the legal problems of noise ordinance enforcement difficult?

4. Two shrill sounds are heard. Sound A registers 90 dBC on a sound-pressure meter while sound B registers 90 dBA. (a) Which sound actually seems louder? (b) Which sound would register higher on a sound-pressure meter?

SUGGESTED READINGS

Numerous excellent textbooks on noise have been written and continue to appear. The following are merely examples of different classes.

1. William Burns, Noise and Man (Philadelphia: Lippincott Company, 1969). One of the more readable general textbooks on the noise problem, including measurement and control.

2. Cyril M. Harris (ed.), Handbook of Noise Control (New York: McGraw-Hill Book Company, 1957). Dated but still the accepted engineering handbook on noise abatement.

3. A.P.G. Peterson and E. E. Gross, Jr., Handbook of Noise Measurement, 6th edition (West Concord, Mass.: General Radio, 1967). The accepted handbook in the measurement field.

4. Bruce L. Welch and A. S. Welch (eds.), Physiological Effects of Noise (New York: Plenum Press, 1970). An excellent collection of papers originally presented at an international symposium on the Extra-Auditory Physiological Effects of Audible Sound.

5. American Speech and Hearing Association, Noise as a Public Hazard, Report No. 4 (1969). An exceptionally useful collection of information about noise. Scope is broad and technical quality is excellent.

6. Karl D. Kryter, The Effects of Noise on Man (New York: Academic Press, 1970). A rambling, sometimes weak collection of notes by one of the pioneers in the study of noise. Considered a standard reference.

7. U.S. Department of Commerce, The Noise About Us, a report of the Panel on Noise Abatement to the Commerce Technical Advisory Board (September, 1970). A relatively comprehensive review of the noise pollution situation within the United States.

Chapter 11

SIGNIFICANCE OF SOLID WASTES

> "Socrates said, 'Those who want
> fewest things are nearest to
> the gods.'" Diogenes Laertius
> (c. 200 A.D.)

The Root of the Problem

Definition

Solid wastes are man's unwanted materials that cannot
flow directly into streams or rise immediately into the air.
They are the nonliquid, nongaseous residue of our manufacturing, construction, cooking, recreation, agriculture, and
other activities that use and then discard materials. Included are outdated newspapers, glass bottles, metal cans,
paper cups, plastic bottles, abandoned automobiles, demoli-
tion rubble, mine tailings, dead animals, flyash, dewatered
sewage sludge, and the garbage from our dining tables.
These are produced wherever man is found--farms, mines,
stores, offices, factories, homes, hospitals, streets, and
even the primitive encampments of traditional nomads.

Determinants of Wastes

When does something cease to be a useful object and
become waste? The answer is rarely clear but is related
to what we use and how we use it. The bone in a roast eaten
for dinner may be cast into a garbage pail as useless to the
owner. But under other circumstances, the bone may be preserved for soup and then presented to an appreciative dog.
You may use a piece of paper for notes, then deposit it in
a waste basket as trash. Later, when lacking a convenient
scrap of paper for further notes, you may fish the paper out
of the waste basket and use the reverse side, thus trans-
ferring that paper back into the category of useful objects.

In other words, the kinds and quantities of solid
wastes today are largely determined by cultural habits and
economic institutions. The technologies exist to recycle

almost any object being discarded, but this use of technology has a price. We frequently are unwilling to pay it. In less technologically advanced societies today, the solid waste problem is almost nonexistent. The Indian peasant, for instance, buys little and can view every object coming into his possession as a continuing economic resource. Cans are salvaged for containers. Papers are used, reused, and eventually burned for heat. Solid body wastes are collected for agriculture. A necessary feature of the affluent society, though, is a high value on human labor and the sustaining of mass production. Manufactured products cannot be inexpensively reused at the place of discard, and collection and processing costs typically become unacceptable if we try to return items back through our retail marketing system.

Objections to Solid Wastes

Objections to the growing piles of solid wastes in our society commonly fall into five categories: public health, aesthetics, occupation of space, collection-processing costs, and degradation of natural resources. All represent forms of economic costs since they will now or later detract from our opportunities to enjoy life. Yet only the costs of collection and processing can be encompassed by our usual economic indicators.

Public Health Hazard

Solid wastes that are mishandled--or not handled--can harbor disease-carrying agents, become air and water pollutants, and pose serious safety hazards both for the general public and for professionals engaged in waste collection and processing. Although food wastes represent only about 10% by weight of all solid wastes collected by municipalities, they offer an attractive food source for insects and rodents, which can set up housekeeping in the remaining 90% of the wastes. The Public Health Service has identified twenty-two human diseases that can be associated with solid wastes. Examples include typhoid fever, cholera, various diarrheas, dysentery, anthrax, trachoma, plague, and trichinosis. Trichinosis, a parasitic infection of humans transmitted by pork, was especially prevalent in the United States until laws began prohibiting the feeding of unsterilized garbage to hogs. (Yet this requirement also had the unfortunate effect of reducing still another ecologically attractive means of recycling food wastes.)

Yet, like the linkage between smoking and lung cancer, the relationships between health and a pile of refuse are often difficult to communicate convincingly to the public or to a hard-pressed city council trying to pare every last dollar from their sanitation department's budget. We may know that a particular disease is associated with specific carriers, such as fleas, ticks, mites, flies, rats, or cockroaches, but rarely can one say with certainty that a person who has just died received his fatal illness from an insect that was nurtured by a known pile of garbage. An occasional exception is noted. For example, murine typhus is rare in Wisconsin so the Public Health Service was able to trace with reasonable confidence the infection of a 33-year-old man in LaCrosse, Wisconsin, to flea bites occurring while he scavanged in a rat-infested city dump.

The affinity of some disease carriers for garbage is legendary. Flies, for example, have been known to climb through five feet of uncompacted earth to reach garbage (although six inches of compacted--pressed--earth would have typically stopped them) and their offspring have successfully returned to man's habitat. Rats are the customary mark of improperly handled wastes. In slum areas, the most successful rat-eradication programs have been where house-to-house searches have eliminated all rat access to solid wastes, their major source of food and shelter.

When mismanaged, solid wastes will also contribute to other forms of pollution that are health hazards. Rain falling on a garbage dump will wash salts and organic material into nearby streams or into ground water, a process known as leaching. Leaching from mine tailings can be particularly dangerous if copper, arsenic, and similar toxic materials are present. Smoke from burning dumps and coal culms (heaps) was considered a sufficient hazard and nuisance to justify passing the first significant federal solid waste program as an amendment to the Clean Air Act of 1965. Even when solid wastes are not burned, the decaying organic material can generate methane gas that causes disagreeable odors, blackens the paint on nearby buildings, and becomes explosive if confined in a nearby cellar or other enclosed space.

The most substantial evidence incriminating solid wastes as health hazards are the health records of sanitation workers, who experience a disproportionate rate of occupational diseases and injuries. A study of the Department of Sanitation in New York City found that muscle and tendon diseases (particularly in the back), cardiovascular disease, arthritis, skin diseases, and hernias could be considered common occupational ailments. The injury rate exceeded all occupations previously studied, except for logging. The rate was more than twice as high as for firemen and policemen and surpassed that of stevedores. In a survey by California of privately-operated refuse collection workers, the disabling injury rate was found to be 180.9 per 1,000 employees, nearly six times the average rate for all California industries. The collection industry's manual workmen's compensation premium rate was $8.05 per $100 payroll in 1967, four times the overall rate for manual classifications.

Accidents with solid wastes are also a concern for the general public. Many municipalities have passed laws requiring doors to be removed from junked refrigerators so children cannot be trapped inside. In the Northeast, injuries to children in broken hulks of abandoned automobiles have become a persistent cause of trips to the hospital. And while refuse remains in the homes, it is a fire hazard as well as a potential cause for falls, cuts, and similar mishaps.

Aesthetic Objections

Solid wastes from private individuals, municipalities, and established industries are a visible and durable blight upon our landscape. Junked automobiles litter streets and fields. Trash and garbage are sprinkled along highways. And open dumps used by municipalities contrary to accepted rules of public health and aesthetics provide eyesores and odors for nearby citizens. The problem has numerous cultural dimensions. For many persons, the wastes represent untidy housekeeping by civilization. Wastes are "out of place," a form of scenic pollution. Mental associations are also doubtlessly involved. To passing motorists, for instance, the wrecked and abandoned automobile may be an unwelcome reminder of material and human mortality.

The cultural aspect is evident when international comparisons are made. One recurrent story recounts the response of a Scottish (sometimes Swiss) bus driver when an American passenger threw an empty cigarette package out the window. The bus stopped and the passenger had to retrieve the package before the driver would continue the journey. On the other hand, some parks in Tokyo on Monday morning

222

can appear paved with litter from weekend picnickers and strollers.

Outside of urban areas, the industries that log the forests and mine the minerals have been among the aesthetically most ruthless. Whether the miners are seeking coal, iron, copper, or another of about eighty different minerals, waste materials known as tailings must be discarded, and these are presently accumulating at a rate of about 1.1 billion tons per year, or thirty pounds per American each day. Eight mining industries--copper, iron, bituminous coal, phosphate rock, lead, zinc, alumina, and anthracite coal--are responsible for about 80% of the tonnage. Some can be used for constructive purposes, such as aggregate in concrete, but most must be piled in vast, ugly, and barren heaps where the wind can blow the dust and rain can wash silt and leachate. Even some of the constructive uses of mining wastes prove disastrous. Radioactive tailings from uranium mines in the Southwest have been used as foundation and fill material for homes but have subsequently been pronounced health hazards because of excessive radiation.

About 800 culm banks (tailings from coal mines) with 10,000 tons or more of coal dust and waste, many near population centers of Pennsylvania, West Virginia, and Ohio, are potential landslide and fire hazards. In 1964, 495 burning culm banks were identified in fifteen states, and the noxious fumes were destroying vegetation, corroding buildings, and--presumably--killing human lung tissue. The possibility of a mud and rock slide similar to the one that struck an occupied school in Aberfan, Wales, is a threat to some nearby property.

Near many large cities, sand and gravel excavations have left raw, gaping holes. But, when this book was being prepared, no industry posed as severe an aesthetic threat to the rural landscape as strip and surface coal mining. Using bulldozers and crane shovels, thousands of tons of vegetation, soil, and rocks--known as spoil or overburden-- must be stripped from the top of a coal seam and pushed aside or strewn down a hillside so the coal below can be excavated unhindered. Unless the top soil is carefully replaced, vegetation will usually not grow on the sterile, rough hillsides for decades. Valleys are submerged under silt. Acids are washed from the spoil and remaining traces of coal to make many streams totally lifeless. About 20,000 active surface and strip mines have been threatening about 153,000 acres of land annually, and the rate is increasing.

In 1965, an area of about five million acres--roughly equivalent to the size of the State of Massachusetts--was estimated to have been devastated by all forms of mineral and fossil fuel mining and processing. By 1980, an equal or greater area will have been defaced by strip and surface coal mining alone unless currently debated legislation successfully curtails abuses by the industry.

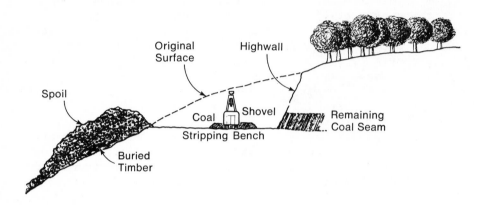

Figure 11-1.   A typical cross-section diagram
of a strip mine operation.

Occupation of Space

Besides being unsightly, wastes occupy valuable space. Trash cans crowd sidewalks. Junked cars prevent parking. In stores and factories where every square foot of floor space has a recognized rental value, the presence of trash containers--whether full or not--adds to the overhead costs of the establishment. And in the home, wastes tend to accumulate most rapidly in the kitchen where convenient space is constantly needed for supplies and equipment. To cope with this pressure for space, we typically move wastes around the home two or three times before a pick-up point for collection is reached.

Where lands are used for waste landfills, the market value of the surrounding property is typically reduced. And subsidence and gas generation on the site can hamper its usefulness for years after abandonment. Furthermore, there are the administrative and human costs for the thousands of public officials who must hire labor to move waste, negotiate disposal sites, and soothe indignant citizens.

Collection-Processing Costs

Throughout the United States, collection and disposal
of solid wastes usually represent the third largest expense
on municipal budgets, led only by costs of schools and roads.
A Public Health Service survey in 1968 indicated that the
average municipal cost just for collection was then about
$5.40 per person annually. Disposal of their wastes cost
another $1.40. When other local and state costs, such as
bond issues, were added, the total bill in the United States
was about $4.5 billion annually.

Degradation of Natural Resources

The earth's prime mineral deposits are limited, and
lower grade ores require proportionately greater amounts of
energy, capital investment, and, often, disturbance of the
environment. In a broad economic context, concern has been
growing about the long-term reasonableness of our market
accounting system that applies only current development
costs to our use of depletable natural resources, such as
petroleum, copper, mercury, and phosphates. In future
years, the argument goes, these resources will still be
required but their recovery will be immensely more costly
because they will be distributed in low concentrations and
difficult chemical forms. Our descendents will suffer from
our squandering habits. Thus, wastes should be minimized
by regarding all depletable resources as more valuable than
the market price would indicate, and we should recycle
wastes whenever the cost is reasonable.

Several contrary arguments are given. Except in a
nuclear reaction, matter is never destroyed; and we should
not exclude the future mining of today's waste dumps, the
sea, and other new resting places. Also, we are constantly
finding substitutes for materials in common use; and, outside of agriculture, few mineral resources are truly irreplaceable in possible future use. For many minerals, low-
grade ore deposits do exist in ample number, although adequate, nonpolluting energy sources may be a limitation.

The debate becomes more manageable when specific resources are considered. A few rarer minerals, such as
mercury, may require a form of rationing. Blatant cases
of malpricing, including the "depletion allowance" on minerals, will certainly have to be revised. Further restrictions on the use of some minerals, notably phosphates, can

225

be partially supported as pollution control measures. But the vast number of technological and social unknowns makes widespread application of a strict conservation policy difficult for the immediate future.

Quantities of Solid Waste

Sources of Waste

If all the solid waste discarded in the United States during the late sixties had been weighed and distributed equally, an average citizen would have received about 100 pounds each day. By the end of the year, he would have been buried under a nineteen-ton heap of dust, rock, broken lumber, sawdust, rotting garbage, rusting automobile hulks, dead animals, torn paper, metal, bedding, and other debris. If the hapless citizen had been able to analyze this miniature mountain, he would have found that about 58% had originated on farms, and most was associated with animal raising and processing.

The next largest category was mining, which accounted for about 31%. Still another 3% came from industrial plants, and only 7% was from urban centers. Half of this, or 3.5%, could be traced to private residences, while the remainder was divided between 2.3% from commercial establishments and 1.2% from municipal government.

Of course, not all of these wastes are equally obvious and objectionable to the average U.S. citizen. Figures from the 1967 survey included the excreta of livestock, the offal of slaughter houses, the stalks, husks, and other vegetable wastes from grain fields, the tailings of mines, the slag of steel mills, and similar materials discarded far from the centers of population. Some, such as the wastes left in the farmers' fields, did not pose a collection or disposal problem. And included were all industries and the large commercial establishments that are usually required by local laws to cope with their own collection and disposal of wastes, often by private contractor. So, unless the industries or commercial concerns were located in urban centers, their waste problems normally did not directly offend the average citizen except as an occasional distant cause of air, water, or scenic pollution.

Table 11-1. Major Sources of Solid Waste in 1967

Source	Solid Waste Generated		
	Million Tons per Year		Percentage
Urban	256		7.0
Residential		128	3.5
Municipal		44	1.2
Commercial		84	2.3
Industrial	110		3.0
Agricultural	2,115		58.0
Plants		552	15.0
Animals		1,563	43.0
Mineral	1,126		30.8
Federal	43		1.2
Total	3,650		100.0

Source: Ad Hoc Group, Office of Science and Technology,
Solid Waste Management (Washington, D.C.: U.S.
Government Printing Office, 1969), p. 7.

A citizen or public agency is more apt to be directly concerned with the seven to twelve pounds per person of wastes originating daily in the crowded urban areas. Here the wastes are a direct nuisance, and the cost of satisfactory collection and disposal are immense. The largest category--from two to five pounds per person daily--comes from private residences; but, since these wastes are usually difficult to distinguish from those of many small commercial establishments, a combined figure of three to seven pounds per person daily is often given.

Table 11-2. Average Solid Wastes Collected in 1968 from Urban Areas in the United States

Category	Lbs./capita/day
Combined household & commercial refuse	4.29
Industrial refuse	1.90
Institutional refuse	0.16
Demolition and construction debris	0.72
Street and alley cleanings	0.25
Tree and landscaping refuse	0.18
Park and beach refuse	0.15
Catch basin refuse	0.04
Sewage treatment plant solids (sludge)	0.50
Total solid wastes collected	8.19

Source: 1968 National Survey of Community Solid Waste Practices (HEW).

Industrial wastes are as varied as the products we wear, eat, and otherwise use in our daily lives. They can include bales of paper, metals, rags, and rubber; piles of slag, sludge, broken glass, and metal scraps; and heaps of food wastes and plastics. Many workers discard as much paper and food scraps during their hours in factories as they do at home. Excluded from this list should be the liquid paints, dyes, plating and pickling liquors, solvents, oils, and adhesives that are considered in the liquid-waste category.

The institutional refuse includes wastes from schools, municipal centers, and similar public buildings, but hospitals represent a special problem since part of their wastes

include the numerous disposable items used today in the
treatment of patients as well as human wastes of operating
rooms and laboratories. These must be handled and processed, usually by incineration, with special precautions.

Municipal wastes include the litter and grit removed
by street sweepers, the grass clippings and fall leaves
from parks, debris taken from the catch basins of sewer
systems, and the sludge of sewage treatment plants. The
proportion of each can vary with the season. Grit is a
winter problem in northern communities that use sand on
highways while leaves overwhelm collection facilities during
the autumn. Both categories are affected by the urban location and socio-economic characteristics of the community,
the landscaping wastes being more of a nuisance in a suburban area than in the inner city.

Households also differ in their waste disposal habits
according to their income, cultural background, age, and a
variety of other factors. One household may generate twice
as much paper as another or triple the food refuse per person. In one house, wood and cloth may be a common disposable item; another household may reuse these materials many
times. Typically, though, paper (cardboard boxes, newspapers, magazines, mail, food cartons, tissue paper, and other
paper forms) account for over half of the total weight in
household waste collections. Glass, china, and pottery

Table 11-3. Typical Composition of Household Refuse

Component	Percent of Total	Pounds per Person Daily
Garbage	12	0.20
Paper	50	1.75
Wood	2	0.07
Cloth	2	0.07
Rubber and leather	2	0.08
Garden waste	9	0.31
Metals	8	0.28
Plastics	1	0.04
Ceramics and glass	7	0.24
Nonclassified	7	0.24
Totals	100%	3.50 lbs.

Source: Based on 1967 analyses in Santa Clara County, Calif.

supply another 10%, about the same as metals. Garbage, the food wastes, rarely exceeds 25% of the total weight and is frequently half that amount. Though they are bulky and obvious, the light weight of plastics generally holds their proportion to less than 5%.

Future Increases

The overall solid waste problem has worsened rapidly in recent years because of at least three major trends: (1) the increase in the number and variety of products to be thrown away, (2) the increase in the amounts of waste--especially packaging--associated with each item produced, and (3) the tendency to cluster certain waste-producing operations in a manner that overwhelms any natural disposal process. Yet there are also countertrends as awareness of the problem grows, our productive institutions change, and technologies become more sophisticated.

Increase in Number and Variety
of Potential Waste Items

Using the Gross National Product as a rough indication of our increased productive capacity, the American economy has averaged 3.8% annual growth between 1950 and 1970. Acquisition of consumer goods, such as automobiles, appliances, and magazines, rose between 4 and 6% annually; and some authorities believe that 4% annual increase in solid waste production is a conservative estimate. One recent study predicted that packaging materials alone will account for a 3.6% annual increase to 1976. At the same time, the variety of goods became greater. For example, the number of major plastic groups listed in the Chemical Economics Handbook rose from nine in 1950 to twenty-two in 1969, and each group had numerous forms and grades. Every group could be characterized by special chemical properties that affect its destructibility and the products formed when chemical degradation occurs.

Accelerated Obsolescence

Just enumerating numbers or even values of products manufactured in the United States does not accurately reflect the amount of wastes being generated. Nonreturnable bottles have been substituted for the previously redeemable

230

beer, milk, and soft-drink bottles. Groceries, especially fruits and vegetables, are now more frequently wrapped. Appliances are encased in multiple layers of corrugated paper that are worthless to the buyer as soon as the appliance is ready for use.

Built-in obsolescence also means that many appliances, automobiles, and other "durable" goods being constructed will replace, not supplement, existing items. Thus, the older item will move more rapidly than otherwise into the waste category.

Concentration of Waste-Producing Activities

The efficiency advantages of economies of scale--the concentrating of economic processes that can share the same technologies and other scarce factors of production--have caused some unforeseen waste problems, especially in the food and related industries. "Convenience foods" means that waste parts of food items, such as vegetables, are now removed in a factory unit where they accumulate and must be reintegrated with the environment at a cost borne without direct reward to the factory operators.

For example, farmers have found that animals can be raised more economically by confining them in sheds or feedlots, often located near urban centers so transportation costs can be minimized. Commercial feedlots were almost unknown in Kansas during the early 1950s. Now more cattle are fed in feedlots than on farms. Since a beef steer produces from twelve to eighteen times the manure of a human, a 10,000-head feedlot is equivalent to a city of 120,000 to 180,000 persons and has to dispose of about 260 tons (fifteen tons dry) of wet manure daily. Numerous states, including Missouri and New Jersey, have called conferences in recent years to discuss techniques for coping with this problem, although rising cost of feed grains may cause a countertrend.

Countertrends

Cries that we will suffer an interminable increase in solid wastes can be viewed with skepticism. Resource shortages, new industrial processes, and enlightened managerial attitudes can change situations quickly. In 1965, the sawmill industry was estimated to be responsible for sixty-five

billion pounds of the entire 182 billion pounds of wastes
produced by all major industries. But new processes now
turn much of this waste into saleable products, and almost
all of the sawmill wastes are expected to be eliminated by
1975.

In the field of packaging, the economy can be described
as having undergone a packaging revolution, but most items
that can be covered are now wrapped. Packaging is expensive
to produce and ship, and the trend may be more towards making packaging lighter, less bulky, and possibly more useful
after its initial function has been served. Consumer organizations are also mounting campaigns against both excessive
packaging and nonbiodegradable waste. And the plights of
local sanitation departments are educating the public to
the costs they pay when manufacturers ignore the disposal
problems of their products. With continued public pressure,
more refined technologies, and changes of attitudes within
corporate management, the eventual control of solid waste
increases to a rate equal or less than average economic
growth appears possible.

QUESTIONS FOR DISCUSSION

1. What would be the possible water-pollution effects ex-
    pected by leaching from a sanitary landfill into a soil
    of fine loam? Coarse gravel?

2. Twenty local workers in a high unemployment district
    have just been laid off because you closed a local strip
    mine coal operation after the operator stated he could
    not afford to restore the stripped land. Making neces-
    sary assumptions, justify your action.

3. Using the wastes that you remember being discarded in
    your home during the past twenty-four hours, how many
    items could you have restored in one form or another to
    some useful purpose? How? Would the effort have seemed
    worthwhile in any case?

4. Using the daily per capita waste figures given in this
    chapter, what average daily waste flow would you expect
    from your local community? Assuming trucks carrying an
    average of twenty tons, how many truck loads would this
    represent? How would you expect this flow to vary by
    week? By season?

5. Discuss in depth the means that the modern manufacturing corporation has in influencing the quantities of solid wastes in our society.

6. Since city councils typically react more favorably to quantitative arguments than simply qualitative, what measurements (indicators) would you consider presenting in a report that describes the seriousness of the public health, aesthetic, and occupation-of-space objections to mismanaged solid wastes?

SUGGESTED READINGS

The standard references in solid waste management are generally published by the Institute for Solid Wastes of the American Public Works Association. Among the more general are two volumes:

1. Institute for Solid Wastes, APWA, Municipal Refuse Disposal (Chicago: Public Administration Service, 1970).

2. _____, Refuse Collection Practice (Chicago: Public Administration Service, 1966).

For a concise, knowledgeable evaluation of the national solid waste problem, refer to the report prepared under the leadership of Rolf Eliassen for the President's Office of Science and Technology:

3. Solid Waste Management: A Comprehensive Assessment of Solid Waste Problems, Practices and Needs, a report prepared by Ad Hoc Group for the Office of Science and Technology, Office of the President (May, 1969). Available from the U.S. Government Printing Office for $1.25.

A more popularized and less accurate description of the problem is:

4. William E. Small, Third Pollution: The National Problem of Solid Waste Disposal (New York: Praeger Publishers, 1970).

Anyone studying the problem in depth should refer to the U.S. Environmental Protection Agency's publications. A free booklet listing publications in print can be obtained by writing: Solid Waste Publications Distribution, U.S. Environmental Protection Agency, Cincinnati, Ohio 45268.

For technical articles on current technologies and management approaches, a standard source is the monthly journal,
Public Works.

Chapter 12

MUNICIPAL SOLID WASTE MANAGEMENT

". . . a large number of reports were received setting
forth that not only were the
streets not properly swept, but
that in many parts of the city,
especially those inhabited by
the poorer classes, the accumulations of days, and even weeks,
remained unremoved. These accumulations consisted not only
of ordinary sweepings, but
mainly of house garbage--putre-
factive animal and vegetable
matter. . . ." Metropolitan
(New York City) Board of Health,
Annual Report (1866)

Current Situation

Typical System

Our overall system for removing discarded materials
from our dwellings and offices is crudely rudimentary compared to the sophisticated manufacturing and marketing system that produces the items for our use. In the usual case,
household wastes are emptied from container to container
until they reach a trash can outside the building. Once or
twice a week, two or three men laboriously lift and empty
the twenty- or thirty-gallon metal cans into the garbage
truck's rear hopper. A compactor--a huge squeezing device--
removes the wastes from the hopper and compresses them until
about 500 to 600 pounds of material occupies each cubic yard
of space instead of the original 100 to 200 pounds. When
the truck is full, it must navigate city streets until a
disposal site or transfer station is reached. Wastes eventually reach the city's dump, which may be called a "landfill" but rarely has the characteristics needed for designa-
tion as a sanitary landfill.

As explained in the previous chapter, many objections have been raised to this system. Natural resources are used once and then discarded, presumably forever. Intense human manpower is focused upon a task that, with some innovations, could be more safely and efficiently performed by machines. Additional administrative and fiscal burdens are placed upon overworked city administrators. Streets are congested by the trucks, and the cans cannot be termed a gracious contribution to urban beauty. Furthermore, trash spills on streets and the din of trucks and cans add to the urban noise.

If a dump is used, the community is constantly threatened by outbreaks of fire with the attendant smoke and odors. Rodents thrive, and valuable land is often wasted as neighboring property values are depressed. If an incinerator is used, disposal costs can be three or five times that of a sanitary landfill; and the curse of additional air pollution is usually added.

New technologies to improve the situation are gradually being developed. In some waste categories, such as commercial waste, these are being applied effectively. But many observers believe that the situation will not drastically change until (a) a diverse set of factors, especially lifestyles and costs for material goods, change and (b) city administrations are prepared to pay more and incur risks to introduce the needed innovations in solid waste handling.

On-Site Processing

Waste Minimization

Each successive step in handling a particular waste represents additional cost to someone, so a logical rule in enumerating possible actions is to begin where the wastes first appear. Unhappily, the situation in our complex society does not produce simple solutions. Those who produce the wastes are often not those who must pay for the disposal. For most intermediaries it is cheaper to pay for handling and to pass the disposal costs to another party. From a total society viewpoint, though, costs are still accumulated. Thus, a careful analysis of costs and benefits must be accompanied by identification of the persons or institutions that pay the costs and reap the benefits.

The first opportunity to reduce wastes occurs before the object is manufactured. Two approaches can be taken: (1) material minimization and (2) object durability. Assuming the object is really needed, it can be designed to provide the functional capabilities with minimum material weight and bulk. Excessive packaging can be avoided. Low-benefit frills can be eliminated. Obviously, though, a manufacturer will use excessive packaging if he is liable for any damage to a product but can pass packaging and shipping costs to a customer. And if low-benefit frills to an object will give a manufacturer the competitive edge, the frills will be added. In neither case would manufacturers have to pay for the added disposal costs.

A second approach would be to extend the useful life of a product. For example, a refrigerator operating for twelve years instead of eight will mean that one-third fewer refrigerators will be added to waste piles during that period. Manufacturers, of course, cannot be expected to display strong enthusiasm for this because fewer products will be sold. In part, too, sympathetic consumer attitudes are also needed since someone must be willing to pay for a more durable appliance and take necessary care of it.

On-Site Recycling

In waste minimization, the material is not manufactured so it cannot become waste. In on-site recycling, the material exists but it can be treated and then reused. This is true recycling with the item being reused for the same function with equal value. A drinking glass is washed and recycled. Dirty clothes are cleaned and recycled for the original function. Newspapers are collected, processed, and reused as newsprint. In degraded recycling or resource recovery, though, the materials are not restored to their original form but are used for a less valuable function. Newspapers are burned for their energy content. Table scraps are added to a compost pile as a future source of nutrients and humus for a garden.

On-Site Disposal

Ideally, on-site disposal would be encouraged. Further handling would not be necessary, and, presumably, further costs would be incurred. But "disposal" is a misnomer. Wastes do not disappear; they are only reintegrated into

the natural environment--soil, air, or water--or are recycled for human use. Because of the threat from water
pollution, only air and land are typically utilized as
waste disposal media. Because land disposal is obviously
impractical for an apartment dweller, partial disposal by
burning wastes and then placing the ashes outside for city
collection has occasionally been encouraged but has usually
produced disastrous air pollution conditions.

New York City, where the American Public Works Association estimated in 1970 that on-site incinerators in multi-
family dwellings could save about $18,000,000 annually,
provided a typical case example after 1951 when an ordinance
was passed to require all new major construction to install
incinerators. By 1967, about 12,000 of these incinerators
were burning roughly 3,000 tons of waste daily, but they
were also producing at least 15% of the City's air pollution
particulates compared to the municipal incinerators' 20%
from 7,000 tons of refuse. And most of the municipal incinerators were obsolete with only rudimentary air-pollution
controls. In the mid-sixties, New York City began a rever-
sal of policies and no new on-site incinerators are now
allowed. Old ones must be either upgraded with sophisticated air-pollution equipment or abandoned.

Technically, the small incinerators pose a harsh technological dilemma. If high combustion temperatures are used
to thoroughly burn the waste, the inexpensive steel chambers
are quickly ruined and auxiliary fuels, usually oil and gas,
must be amply supplied. But lower combustion temperatures
produce obnoxious particulates unless expensive and delicate
control mechanisms are added. Furthermore, while an ordinance can require proper maintenance, most cities find enforcement impossible because of scattered locations.

Collection and Transportation

Multiple-Method Approach

If waste disposal cannot occur on-site, the wastes must
be transported elsewhere; and, if a truck or other container
is being used, this step must be preceded by collection, the
assembling of sufficient wastes to make the trip economically bearable. Though the truck remains the primary means of
collection and transportation, numerous other methods are
beginning to emerge. Garbage grinders installed in the

waste-water system remove kitchen wastes. Newspapers are increasingly being collected separately for recycling. Pneumatic systems, dry pipes that use air pressure to provide a propelling force, are being installed for collection. Large steel containers that are periodically removed for emptying are simplifying collection and transportation for large commercial centers, such as shopping malls. And even the familiar compactor truck system is undergoing an evolution with plastic bags, automatic loading equipment, and multiple-container networks that will gradually produce a more economic and effective means of moving wastes to a disposal point.

Garbage Grinding

The grinding of garbage in a small unit beneath sinks has been quietly accepted in most cities of the country. Wastes are hydraulically flushed through the sewer system, a waste transport tube that already exists but is usually underutilized, to a sewage treatment plant where solids are removed mechanically and, being predominantly finely ground organic materials, can be easily decomposed with the sludge.

Although the addition of garbage to sewage can double the amount of solids being carried, sewage is already 99.9% water so doubling the solids will not make a noticeable difference in the total volume passing through sewers. In some isolated cases where sewer systems have been poorly constructed or maintained, the garbage has caused clogging but the typical experience has been that garbage improves the performance of the sewers by scrubbing grease and other accumulations from the sewer walls. And, because the treatment-plant design is generally based on water flow, only a few plant components--those handling solids and the sludge digestion--have to be enlarged.

The American Public Works Association estimated in 1970 that garbage grinding typically adds between $.50 to $1.50 to the annual per capita cost of operating sewage treatment plants. Including cost of grinder, installation, electricity, sewage treatment, and all other costs, the average household probably pays about $35 to $45 per ton for this collection and disposal system. Balanced against this cost is the elimination of putrescible materials from wastes with a corresponding reduction of health hazards, odors, insect infestations, and other problems associated with garbage standing in houses or on streets. In institutions and commercial establishments, such as schools and restaurants, the advantages of grinding are even more compelling.

Hydraulic waste systems that can transport all wastes have been installed in a few institutions, such as hospitals, but these have generally been abandoned because of problems with grinders, removal of the waste from water without a complete treatment plant, and the problem of disposal when glass and metals have become dispersed throughout water-soaked paper. Meanwhile, efforts to develop a citywide system have continued. One company that specializes in paper-making machinery has installed an experimental system that will allow recovery of recyclable materials, including the sewage-soaked waste paper fibers, but reliability and the economics have been questioned.

The use of dry pneumatic systems for collection of wastes has also been tried on a limited scale. About six systems are operating in Sweden, one serving a community of more than 5,000 dwellings. In the United States, one hospital, some apartment complexes, and Disney World in Florida also have installations. Untrained personnel can empty wastes into building chutes, and these are brought through eighteen-inch steel pipes to a central bin. The cost is high, however. One million dollars is not unusual for an institution with scattered buildings though the pipes carry wastes for only a few minutes each day. Furthermore, transportation and disposal problems still remain since the pipes rarely extend more than a fraction of a mile.

Truck-Related Technologies

Most commercial concerns are now using large steel containers that they rent from private contractors. Sizes typically range from three to fifty cubic yards of space. A compactor unit is sometimes installed inside or outside the container to compress the wastes into the space. For larger containers, special trucks simply exchange an empty unit for the filled one and empty the contents at a disposal site. For smaller containers, new compactor trucks are equipped with a special device for lifting the container and emptying it automatically. Some cities have been experimenting with containers located at strategic sites throughout the city.

Substitution of plastic or paper bags for the familiar metal trash can is another growing trend. Without the heavy weight, awkward shape, and sharp edges of the cans, accident rates for collection crews have been cut. Fewer men are needed because they do not tire as quickly, and empty cans do not have to be returned to the curb or yard. Even when the cost of bags--which can be ten cents or more for a thirty-gallon bag--are considered, the cities using plastic bags have generally found them considerably less expensive. Some cities have devised systems that automatically throw the bag into the truck, permitting a single operator to conduct collection. While theoretically advantageous, the breakage of bags and occasionally misplaced bags makes this automatic system work erratically even if curbside automobiles can be successfully cleared.

Other concepts include the "mother" truck and the train collection systems. Essentially, the "mother" truck system involves a large compactor-storage truck that is served by a fleet of small-capacity scooters or pick-up trucks. The larger trucks remain on avenues while the smaller vehicles, each operated by a single man, collect waste on side streets, periodically returning to empty wastes into the larger vehicle. The train system uses a line of connected, wheeled containers pulled by a tractor. Filled containers are exchanged for empty ones at predesignated points along a route while smaller vehicles are assembling the waste.

Transfer Stations

Since at least 90% of the nation's population depends upon some type of landfill for waste disposal, the wastes typically have to be carried for many miles beyond each community's border to find an environmentally acceptable and inexpensive site. Why use an expensive, small-capacity compactor truck with as many as three men for this trip? In most cases, the answer lies in the expense of constructing and operating a transfer station where wastes can be moved to large vans, barges, or trains.

A typical transfer station may cost three dollars per ton for operation; and, with increasing labor costs, the expense will obviously rise further. Yet, cities are having to send the waste further and further from borders, and a transfer station looks increasingly attractive. Also, a transfer station can serve as a center for resource recovery, the recoverable items being sorted and sold near an industrial area. This concept of a multipurpose processing

station is now regarded as one of the more promising possibilities of solid-waste management.

Rail Transport

Rail lines enter every major city; and, except for
barges and pipelines, railroads can provide the lowest
transportation costs for long-distance, high-density cargo.
With the rising land values and traffic congestion near
urban centers, a logical innovation is to transfer the waste
to trains for disposal in more distant and isolated sites,
such as abandoned coal mines. No new technologies are
needed, and projections of estimated costs are often less
than current alternatives. Despite repeated attempts,
though, the concept has floundered on a combination of
institutional conservatism and adverse public reaction.

Risks are involved. To obtain low freight rates, the
wastes must be compressed by a powerful compactor to a density approaching coal. Transfer facilities must be constructed at both ends, and disposal arrangements must be
concluded at the terminus. To protect these investments,
cities have insisted upon long-term, low-cost contracts
with railroads. On the other hand, railroads have been
reluctant to provide such long-term commitments because
they suspect that their costs will rise unpredictably.
Behind this conflict has been vociferous opposition by local
residents of the receiving areas against "receiving someone
else's garbage."

Processing and Disposal

The Problem

When wastes have been collected and transported, there
must be either further processing or immediate disposal.
Sometimes the wastes are simply dumped and labelled "landfill." Less than 6% of 12,000 land disposal sites surveyed
by the U.S. Public Health Service in 1968 were classified
as meeting minimum management standards. Some wastes are
burned, but the survey classified about 75% of the 300 incinerators in the U.S. as inadequate, generally because they
were responsible for severe air pollution. Though governmental controls are tightening, some wastes are legally
dumped in the sea, and still other items--especially mining

and agricultural wastes--are left scattered more or less conspicuously about our urban and rural landscapes.

When alternative methods of processing and disposing of solid wastes are examined, the options generally fall into three broad categories. (1) The material can be processed for reuse, either for a similar product, such as paper reprocessed into paper, or a completely different material, such as compost. (2) Heat can be used to diminish the bulk and weight, some of the materials being transformed into gases. (3) The material can be buried.

Resource Recovery

Resource recovery means the retrieval of materials from wastes for some economically useful function. The term can refer to the complicated sorting of paper, cans, bottles, and other materials from domestic wastes and subsequent processing, or to a steel plant's homogeneous wastes that are returned to the furnace and remelted. Recovery can be for an equivalent use, such as the reprocessing of newsprint (though the term "recycling" is usually applied), or it can be for a degraded use, such as the recovery of heat or compost from domestic wastes.

Materials that--under specific conditions--can be economically separated from wastes, processed, and reused today include paper of many different grades, ferrous metals, glass, aluminum, oils and tars, copper, inert ash and grit, certain chemicals, and a variety of lesser items. Since paper, glass, and metals represent the bulk of residential and commercial wastes, they are receiving highest priority in most recovery efforts. Industries that assemble and redistribute wastes are known by various names, such as paper stock companies and secondary material industries.

In general, these recovery-oriented industries suffer three chronic problems: (1) location, (2) purity, and (3) market fluctuations. In a few instances, the location of a material relative to its markets and reprocessing points are relatively unimportant. Gold recovery is rarely deterred by these considerations, for example. Manilla IBM tabulator cards are worth about $100 per ton on the stock paper market because of their high-quality, long fibers, and consistent composition, so they are worth transporting long distances.

The value of most waste materials, though, is so low that transportation costs for any appreciable distance makes them losers for the seller. Glass cullet (scrap) has traditionally sold for twenty to thirty dollars per ton at a glass plant, but the cost of loading and sending a truck of several tons for even a few miles can surpass that value. Processing costs to produce compost are estimated to vary between fifteen to twenty-five dollars per ton, and the nutrient value of the product for fertilizer is very low. Thus, since wastes are produced in a city and compost is generally applied in the country, transportation provides a critical cost factor since disposal costs in a rural sanitary landfill--the usual alternative to composting--will rarely exceed four dollars per ton though composting can reduce the weight of waste by about half.

Fluctuations in markets prevent industries from making capital investments or safely assembling waste for resale. The price for newspapers in recent years has several times risen from zero to twenty-five dollars per ton and collapsed again within a year. Dealers who buy papers and must store them until car lots are assembled must constantly gamble with heavy losses possible. Glass, steel scrap, and waste oil have also suffered from fluctuations, although not as dramatically as paper.

To make an economic success of recovery, purity of the material is a key factor. The reasons can be compared to the ones that make sewage, which is 99.9% pure drinking water, unwanted for any use. A single shred of plastic coating on a piece of paper being recycled will create a flaw, causing the new paper to tear when being printed. Copper wires interlace the steel body of an automobile, but scrap steel is usually considered unacceptable when the percentage of copper rises above 0.02%. To be sold, most waste products have to be graded according to the needs of various manufacturing processes. Even the slightest deviation, such as a piece of yellow paper in a bale of white, can cause rejection of the entire lot and a costly loss to the dealer. As a consequence, most of the wastes being recycled in the U.S. are by-products of industrial processes, such as steel fabrication or paper envelope manufacturing, where the purity can be assured.

Immense effort has been expended on development of separation devices for mixed refuse. Ferrous metals can be removed by magnets. Bottles and some other heavy items can be removed by ballistic devices, mechanical hammers that knock heavier items from the lighter paper and plastics. The air classifier, a series of vertical, zig-zag tubes with air blowing upwards, has been borrowed from agriculture where beans and similar products have to be separated from shells and husks. Heavier, compact objects fall, while lighter, bulkier objects rise with the air. Fiber optics can be used to scan glass and classify it by color. None of these, though, have demonstrated the discrimination that is needed for the degree of purity required by most industries.

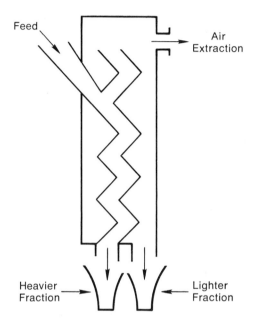

Feed

Air Extraction

Heavier Fraction

Lighter Fraction

Figure 12-1. The air classifier adapted by Stanford Research Institute to waste separation depends upon upward air currents to carry lighter materials to a separate compartment from heavier materials.

What happens if one accepts the composition of mixed wastes but tries to remove obnoxious qualities and then seeks a market for all or any part that can be separated? This has been the concept behind composting. As indicated earlier, though, no market has been found that could support the high cost of processing. The technologies of composting have been highly developed, and at least eighteen composting plants have operated in the U.S. during the past decade. Most have solved the technical problems but have floundered on the market and cost problem. Most of these efforts received municipal subsidies, so municipalities feel a degree of justification in seeking other, less costly alternatives. By contrast, at least twenty-five composting plants serving cities with populations over 20,000 persons were reportedly operating successfully in Europe and the Middle East during the sixties. On a worldwide basis, the number of large municipal composting plants number in the thousands.

Heat Processing

Incineration

Incineration, the burning of wastes under controlled conditions, offers a municipality some important advantages. Waste volume can be decreased by 80 or 90% by a conventional, low-temperature incinerator or as much as 97% by an experimental, high-temperature incinerator. Thus, the transportation costs can be drastically reduced if wastes are incinerated near the collection area and only the ashes taken out of the city. Furthermore, since the ashes are sanitized, they do not pose the environmental hazards of the original wastes. Yet less than 400 municipal incinerators are operating today, and almost an equivalent number have been abandoned in recent years because of air pollution problems and, to a lesser extent, traffic congestion.

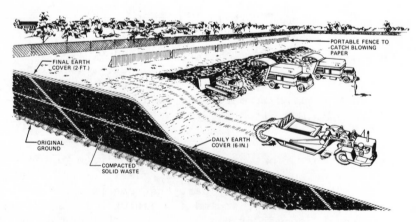

Figure 12-2. Cross section of a sanitary
landfill placed in a ravine.

Source: Sanitary Landfill, New York State
Department of Health.

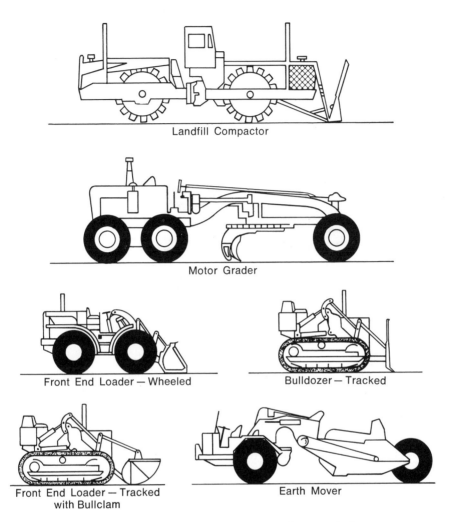

Figure 12-3. Equipment commonly used in
operating a sanitary landfill.

This failure of incinerators to match expectations is
not new. The United States' first major incinerator was
probably constructed in 1885 on Governor's Island in New
York; and, by 1903, two heat-recovery incinerators using
water-cooled walls and dumping grates had been inaugurated
in Manhattan with elaborate fanfare and predictions that
future cities would be heated by incineration. Yet these
plants were quietly abandoned within several years because
of operating difficulties, air pollution, and the inconsistent level of heat available from wastes. Later incinerators

have had improved air control, mechanical stoking, and a continuing list of elaborate air-pollution control devices, but the same complaints continue to be registered.

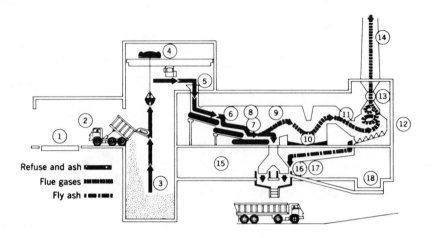

Refuse and ash
Flue gases
Fly ash

1. Scales
2. Tipping floor
3. Storage bin (Pit)
4. Bridge crane
5. Charging hopper
6. Drying grates
7. Burning grates

8. Primary Combustion chamber
9. Secondary combustion chamber
10. Spray chamber
11. Breeching
12. Cyclone dust collector

13. Induced draft fan
14. Stack
15. Garage—storage
16. Ash conveyors
17. Forced draft fan
18. Fly ash settling chamber

Figure 12-4. The interior design of a typical modern municipal incinerator.

Source: National Association of Counties Research Foundation, Solid Waste Management, prepared for the U.S. Public Health Service, Department of HEW, Washington, D.C.

Technically, acceptable incinerators can be constructed but the financial cost has typically been marginal compared to use of sanitary landfills. In 1949, incinerators generally cost about $2,000 for each twenty-four-hour daily ton of installed capacity to burn waste. By 1964, rising labor costs had generally tripled this figure, but few incinerators had adequate air-pollution control equipment. As air-pollution control ordinances became stricter, incinerator costs climbed rapidly. In 1970, two incinerators—Chicago and Montreal—were completed for about $12,000 per daily ton of capacity, and a preliminary estimate for the cost of constructing an incinerator in the former Brooklyn navy yard exceeded $30,000 per daily ton of capacity.

These new plants included heat-recovery equipment since, if a market existed for the steam, the economics had shifted in favor of heat-recovery designs. Because water removes excess heat from the furnaces, a water-cooled furnace does not require as much air to be blown into the combustion chamber as for the older, refractory-walled incinerators. Thus, the air-pollution equipment does not have to be designed for as large capacities of emission gasses. Also, the higher temperatures that are possible produce improved combustion.

The cost of constructing an incinerator, though, is only part of the total cost. Operating expenses can be high. Skilled operating engineers are needed. Maintenance for the elaborate equipment is a necessity. Custodial personnel are needed. Typical operating expenses (excluding capital costs) where statistics are relatively reliable have ranged from five through twelve dollars per ton of refuse processed in large municipal plants. In 1968, Philadelphia (Northwest) reported an operating cost of $6.16 per ton of refuse; Chicago (average of three plants), $5.87; and Milwaukee, $6.04. Despite these costs, many plants--including Chicago's new heat-recovery incinerator--were still reporting air-pollution problems in the early seventies.

Air pollution from incinerators is an unusually complicated problem because of the wide range of materials being burned. Typical municipal wastes contain virtually every chemical material produced by modern industry. The combination of these materials, high temperatures, and the presence of catalysts inevitably means that toxic gases and particulates will be formed. Some, such as phosgene (tolune diisocyante), are toxic in even trace amounts. Other toxic pollutants that often are detected include oxides of nitrogen, ammonia, alderhydes, esters, compounds of silicone, sodium, potassium and magnesium, and various polynuclear hydrocarbons. Besides posing a threat to human health, these materials are highly corrosive to equipment. Faced with these difficulties, specialists have been exploring new heat-treatment concepts, including the high-temperature incinerator, fluidized-bed incinerator, pyrolosis, and combined pyrolosis-incinerator units.

High-Temperature Incineration

Some of the difficulties with conventional incinerators may be overcome by using considerably higher temperatures that reduce wastes to a harmless grit, which can be used as an aggregate in concrete or foundation material in construction. These high-temperature (about 3,000° F.) plants would eliminate the need for some of the more complex stoking and air-control devices because burning would be more rapid and complete. A new set of problems emerge, however. Equipment and controls would have to be made of special metals to resist the high temperatures, and even these would be strained. Instead of the carbon monoxide of lower combustion temperatures, nitrogen oxides would become a hazard. This type of incineration has been tried only as experiments, and some of these have been disasters with the molten residue "freezing" in the bottom of the furnace.

Fluidized-Bed Incinerator

If air is pumped through the bottom of a vertical furnace, a mass of sand or other particles can be made to float in the furnace, being supported only by the air pressure. By using hot air, the "bed" of particles is heated and, when finely divided particles of waste are sprayed onto the bed, they are engulfed and thoroughly burned. The system becomes self-sustaining with the heat from the burning materials being used to maintain the temperature of the circulating air. Sufficient excess heat is produced that generation of electricity is a possibility. But the fluidized-bed incinerator requires the waste to be shredded, and the fluidized bed can become unstable if the wastes are not distributed evenly. Also, highly skilled manpower is required, and the energy requirements for pumping air and shredding wastes are high.

Pyrolysis

Pyrolysis is the destructive distillation of wastes by heating materials in a closed vessel without oxygen. In historical terms, this is the ancient art of charcoal making. Gases, including methane, ethylene, and ethane, are given off and can be used to heat the chamber, thus making the process self-sustaining. Since different gases are vaporized at different temperatures, the more economically valuable ones can be separated. The residue that remains is mainly charcoal. One plant has been constructed by a rubber company to recycle oils from waste rubber, and pilot pyrolysis units have been tested in several cities for reducing the volume of urban refuse.

In theory, pyrolysis possesses the advantages of incineration plus the potential of controlling almost all gases emitted. In practice, pyrolysis has encountered severe problems. If wastes become water soaked, vast amounts of heat are necessary to dry the wastes before pyrolysis can effectively occur. Materials must be ground finely for the heat to penetrate evenly, and the heating can be a time-consuming process. Also, some units have proven sensitive to sudden changes in composition of entering materials. All require skilled, knowledgeable operators.

Combination Units

A recent and promising trend in experimentation with heat processing has been the combining of several concepts, especially pyrolysis and incineration. One design consists of a two-furnace installation that chars the waste at high temperatures and little oxygen in the first unit, then burns the gases in the second unit. Most of the exotic pollutants are trapped in the char while the gases can be burned at a sufficiently low temperature that few nitrogen oxides are formed.

Another approach is the addition of shredded wastes to coal that is fueling boilers for electrical power plants. In the past, similar programs have failed because of the high cost of shredding, the differences in heat content between various wastes (e.g., wet leaves versus gasoline-soaked rags), and the administrative bother for utility companies that have few incentives for innovation. Furthermore, the heat content of wastes is not high, being more comparable to peat than coal; and a typical city needs considerably more electricity than can be generated from wastes. However, major European cities--Paris, Amsterdam, Dusseldorf, Rotterdam, and Mannheim, among others--have successful heat-recovery incinerators that generate electricity. Key factors are the high cost of coal in Europe, availability of skilled manpower at lower costs, and a more responsive public attitude towards constructive waste disposal. With rising costs of energy in the U.S., however, more heat-recovery efforts will probably occur.

Landfill

Sanitary Landfill

Sanitary landfill is the controlled burial of wastes under compacted earth. Costs are relatively low, generally being between two and four dollars per ton of waste, excluding land cost. Also, the procedure is technically simple so highly trained personnel are not needed.

To construct a sanitary landfill, a well-drained site distant from any body of water should be used unless special precautions are to be taken. Wastes are compacted to a depth of six or eight feet, then covered daily with six inches of compacted earth. When the landfill is completed, it should be covered with at least two feet of compacted earth. To provide a constant supply of earth, several techniques are used. If a hill is available, the landfill can be started at its base and earth pushed daily from the hillside. On level terrain, a trench is dug and the fill is covered with the excavated earth. Some sites need special drainage systems and, if near a populated area, methane gas vents are necessary. In practice, seven problems plague the operation of sanitary landfills.

(1) Inadequate soil coverage and compaction.
(2) Continual generation of methane gas.
(3) Subsidence of land after completion of landfill.
(4) Inappropriate sites.
(5) Inadequate control of blowing refuse, especially paper.
(6) Decrease in land value for other uses.
(7) Opportunities for corruption.

If flies and rodents are to be avoided, the fill must be covered daily with the six inches of compacted soil. Also, decomposition of wastes will cause spontaneous combustion unless the wastes are covered. In the 1968 survey, only 14% of the landfills were being covered daily and 41% were never covered. Open burning and a resulting air-pollution problem existed on about three quarters of the sites investigated.

During the decomposition of the organic wastes, methane gas is generated. If not vented, it may travel considerable distances and has been known to fill cellars of neighboring buildings, creating an explosion hazard and disagreeable odors. Even when the landfill is properly compacted and gas vented, the soil will subside as wastes decompose. Usually about 90% of the subsidance occurs within five years, but the time scale may be considerably longer. Although it is possible to construct buildings on subsiding landfills, special foundations must be used or the building will crack and, possibly, collapse. To avoid this type of problem, most landfills are used for recreational purposes when completed.

Sanitary landfills have a reputation of contributing to water pollution. Bare rock or impervious soils have to be avoided since water will tend to flow along the rock or soil surface. Besides stagnant pools, blowing refuse littering the ground and shrubbery are another mark of an improperly controlled landfill. Scavengers are usually prohibited or carefully controlled since the lifting of refuse can cause it to scatter.

Sanitary landfills have an immense appetite for land close to cities. In theory, it is possible to continue adding "lifts" to the top of the landfill each time it is covered, but this creates a visible nuisance to the surrounding community and requires the stripping of soil from land elsewhere. Only a few cities seem endowed with ideal sanitary landfill sites. Los Angeles County, for example, has an abundance of deep, dry ravines and has been able to program its future sanitary landfill operations for another fifty years.

Corruption can also become a specially acute problem with sanitary landfills because of their isolated location. Truck drivers are tempted to bribe the fee collection clerk, who usually sits in a small shed beside the gate to the landfill area. Cities will rarely invest in automatic weighing equipment or other devices that will keep a record of truck arrivals.

Landfills are not necessarily entirely negative features of our waste-disposal system. In recent years, an increasing effort has been made to identify land, such as gravel pits, that will benefit from a landfill. More imaginative use has been made of the recovered land. In some cities, hills have been constructed for skiing. In others, landscaping has been used to improve the attractiveness of parks.

Coverless Landfills

In 1963, the town of Meax, France, began milling (grinding or chopping into small pieces) wastes for a landfill
without cover. This approach was rapidly adopted across
Western Europe and in several Canadian cities. In 1966,
Madison, Wisconsin, tried the method, and it has been followed elsewhere.

Because milling causes putrescible matter to be divided
finely throughout paper and other wastes, the fill does not
support insects or rodents nor will it blow in the wind.
Digestion of the organic material occurs more quickly than
in an ordinary sanitary landfill, and leaching is not a
problem. While the wastes will sustain fires, the fire
spreads slowly and can be quickly extinguished. Although
health authorities recommend covering the final fill with
earth, this does not have to occur each day. Most communities, though, still find that the cost of soil for cover--
if they want a sanitary landfill--is less than the equipment, manpower, and energy costs of milling.

Conclusion

Perhaps the most severe problem with solid waste disposal is the lack of public support for a proper solution.
Funds for development of clean incinerators have not been
provided. In seeking suitable sites for a sanitary landfill,
a city can rarely move beyond its borders because of cries,
"Don't send us your garbage." When new facilities are built,
their operating budgets are starved, and there seems to be a
universal attitude that employment in the waste-disposal
field is an ideal opportunity for rewarding political
friends regardless of their technical skills.

Perhaps the climate will change with the economic
shifts that are accompanying shortages of materials and
energy sources. Inspired political leadership on the national level could also stimulate the development of needed
technologies. Until then, however, solutions will undoubtedly occur on a limited, uncoordinated basis.

QUESTIONS FOR DISCUSSION

1. If you suspected a fresh water lake was being polluted
    by a sanitary landfill, what tests would you request for
    confirmation?

2. Describe three constructive ways that solid wastes could
    be used by a typical U.S. city and then identify at
    least one argument against each.

3. List three or more steps you would take to organize a
    major community effort to recycle solid wastes and then
    identify two major problems you would expect to encoun-
    ter.

4. In terms of funds spent, which aspect of handling solid
    wastes would appear to offer the greatest opportunity
    for cost savings: collection, transportation, disposal,
    or other?

5. Consider solid waste handling in your community. How
    many features are due to administrative convenience or
    tradition and could probably be advantageously changed?

SUGGESTED READINGS

A generally recognized reference in solid-waste disposal has been published by the Institute for Solid Wastes
of the American Public Works Association:

1. Institute for Solid Wastes, APWA, Municipal Refuse Dis-
    posal (Chicago: Public Administration Service, 1970).

Several authoritative anthologies have been published
by Technomic Publishing Company, Inc., 265 West State Street,
Westport, Conn. 06880. Perhaps the best is:

2. David G. Wilson (ed.), The Treatment and Management of
    Urban Solid Waste (Westport: Technomic, 1972).

As in the previous chapter, an immense amount of literature, much of it highly technical, is available from the
U.S. Government. A complete list of available literature
and its source can be obtained from the:

Solid Waste Information Materials Control Section
U.S. Environmental Protection Agency
Cincinnati, Ohio 45268

You may also be able to obtain some information from
your state government, which can often provide free literature.

Chapter 13

PEST CONTROL

> "He gave also their increase
> unto the caterpillar, and their
> labor unto the locusts."
> Psalms 78.46

Inherent Problems

Definitions

A pest is any insect, mammal, bird, plant, or other
living organism identified as unwanted. But who does the
identifying? Anyone. The family having a picnic in a
meadow will curse the bee buzzing around their lunch even
though the neighboring farmer will describe the same bee as
a beneficial insect. Neither will be considering the bee's
role in the broader ecosystem encompassing the forest,
streams, and the meadow. This conflict in labels was not
important as long as each person was restricted to simple
pest control technologies consisting of fly swatters, traps,
and similar selective devices. Today, however, we possess
lethal chemicals and other pesticides that can infiltrate
total ecosystems and are often not selective by species.

Yet pest control is a necessary evil to maintain modern
civilization. To feed the large human population, a degree
of monoculture--the planting of a single type of grain,
vegetable, fruit, or other plant in one area--must be used.
For opportunistic species of plants, animals, and insects,
explosive growth is inevitable. The response advocated
today by most authorities is of integrated pest control, the
orchestration of techniques and technologies to limit damage
from a specific pest. New farming practices, encouragement
of natural predators and diseases, and numerous other
methods--including chemical controls--minimize danger.

As a policy problem, the wise use of chemical insecticides is only one dimension of the broader issues posed by
economic poisons, all toxic chemicals developed to improve
productive efficiency and decrease losses of property and
resources. Additives for preservation of prepared foods

257

are one dimension. Also included are the fungicides, herbicides, rodenticides, bacteriacides, and similar compounds. Many of these are products of elaborate research and development over the past thirty years; but, in many cases, we do not yet possess adequate information, public awareness, and political sophistication to exercise adequate social controls of their use. This problem will undoubtedly continue to spur legislation in the approaching decades.

Objections to Common Pests

When benefits of pest control are compared to the costs, the benefits can generally be classified into three broad categories: economic, health, and psychological. Some pests, such as rats, are sufficiently objectionable that benefits falling into all three categories are high. Other pests, particularly agricultural, directly bother only one segment of the population and then primarily for economic reasons.

Economic Costs. In 1970, chemical manufacturers produced about one billion pounds of synthetic organic fungicides, herbicides, and insecticides. This represented between 50 and 75% of the world's production, and about half consisted of insecticides and related chemicals. Value approached one billion dollars, about triple the amount ten years earlier.

The primary justification for this expenditure of resources was the prevention of property loss, especially food. No one knows or can even give a credible estimate of the losses in agriculture that can be attributed to insects, rodents, fungi, and other common classes of pests. Before 1945, a figure often used in the U.S. was that about 10% of all crops were destroyed by insects and rodents. No one knows the origin of this figure. When a crop does not attain its potential, it is usually difficult to divide the blame between insects, disease, weeds, rainfall, sunlight, or lack of nutrients in the soil.

In the early 1960s, the U.S. Department of Agriculture estimated that over 15% of the losses in cotton could be attributed to insects and slightly more than 10% to disease. For soy beans and rice, weeds were the primary culprits with losses exceeding 15%. More losses occurred when crops were moved to storage and, finally, to individual homes. Some literature maintains that rats ruin ten times the amount of

258

food that they eat by contamination with droppings, hair, and urine.

Losses in food supply are even more serious in developing countries. In Asia, the acreage treated with pesticides is still proportionately small, but the International Rice Research Institute in the Phillipines has estimated that yields with modern rice strains could be increased between 50 and 100% if insects are scientifically controlled. If a severe leafhopper infestation occurs, as frequently happens, the yields in treated fields can be five times that in untreated. Another estimate for India set the losses at about 50% of all food growth due to rodents, insects, poor handling, and improper storage and distribution. Other estimates place the loss of production in India at about 10% due to insects and rodents in the fields only.

In Africa, the Quelea bird is believed to consume millions of tons of grain each year in the center region of the

Ant

continent. In West Africa, the tsetse fly has posed a continuing though diminishing threat to livestock. In Latin America, the transmission of rabies by vampire bats provides an unusual challenge for pest control. Throughout much of the world, wood-consuming insects, such as termites and beetles, cause millions of dollars damage annually. But, for much of the world, the economic issues are overshadowed by the role of pests as vectors of disease.

Public Health. Pests can affect human health by (1) direct attack and (2) indirect infection as a vector agent. Stinging or biting wasps, bees, ants, flies, mosquitoes, and even rats can cause pain, injury, and death. Spines or hairs of some caterpillars, such as the tussock moth's, cause a rash a and discomfort when they touch the skin. Bites from some scorpions, snakes, and spiders can pose a threat to life, but these pests are rare.

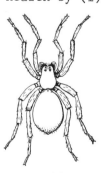
Spider

Honey bee

Scorpion

Historically, pests have probably had their greatest impact on human affairs by the transmission of disease.

About one quarter of Europe's entire population was
estimated to have died during the fourteenth century
from the Black Death, which is assumed to have been
mainly bubonic plague spread by the flea and rat.
Napoleon's retreat from Moscow has been attributed
as much to louse-born typhus as Russia's inhospi-
table weather. Yellow fever has decimated populations in tropical areas and was a recurring terror
in the United States until this century. Today,
malaria is still a severe health burden upon the
developing world, and schistosomiasis--a parasite spread by
freshwater snails--is following the extension of man-made
irrigation systems. But pest-born diseases in the United
States remain a comparatively rare occurrence. Visitors
sometimes arrive with tropical diseases. Ticks can spread
the Rocky Mountain Spotted Fever. Encephalitis can be carried by the Culex mosquito, and the Cone Nose bug can cause
Chagas' disease. Yet many physicians practice for entire
careers without encountering one of these illnesses.

Flea

   Psychological Well-Being. One of the main reasons for
low incidence of insect-born diseases in the United States
is the widespread entomophobia, the fear and hatred of insects. "Bugs" are associated with filth and disease.
Roaches and flies can drive U.S. households into frenzies
of anger and alarm. From another point of view, however,
this same emotional feeling of total warfare against insects
has hampered the rational use of control measures.

   Scope of the Response. Agricultural use of pest control dominated its development. Agricultural research has
provided the major force in developing new approaches to
problems of pest control; and more people, funds, and tonnage of pesticides are used for agriculture than for any
other application. Most of these efforts can be classified
as either chemical or bioenvironmental in basic approach.

Chemical Pesticides

Major Categories

   Chemical pesticides are generally toxic liquids, gases,
or solids that are lethal to pests. The toxicological
mechanism varies immensely between chemicals. Major categories are stomach poisons, contact poisons, fumigants
(smothering agents), and desiccants (drying agents). As

ovicides, larvacides, or adultacides, they can also be directed towards a particular stage in an insect's life cycle.
From an environmental point of view, three characteristics
of chemical pesticides are critical.

(1) Toxicity, the amount of the chemical required
    to cause death or disability in different or-
    ganisms, provides an index of the pesticide's
    strength.

(2) Persistence, the longevity of a chemical's
    toxic properties in the natural environment,
    indicates the threat that the chemical may
    have for organisms outside the target area
    or for future uses of the area treated. Some
    chemicals, such as some chlorinated hydrocar-
    bons, have a relatively short persistence in
    their original form but their metabolites,
    the chemicals that are produced when the
    original chemical disintegrates, are some-
    times more dangerous and persistent than the
    original.

(3) Selectivity, the range of different organisms
    affected by the chemical, provides some guid-
    ance about the general effects that a chemi-
    cal may have when distributed widely. Unfor-
    tunately, the mode of action may vary with
    different species, and this limits our abil-
    ity to identify all of the effects.

Because of the intensive demand for chemical pesticides, nearly 1,000 chemicals are now used in over 32,000
registered pesticide products. Most insecticides, though,
can be classified under four broad chemical categories: the
"natural" chemicals, the organophosphates, the chlorinated
hydrocarbons, and the carbamates. Among herbicides, most
are known as inorganic chemicals, acid derivatives, or
nitrogen compounds. In addition, one should consider a
variety of bacteriacides, and rodenticides.

The "Natural" Chemicals

Historically, a wide variety of household chemicals,
herbs, and common inorganic chemicals have been used for
insect control. The household chemicals are not spectacularly successful but neither are they without effect. Our

Bed bug

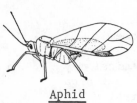

Aphid

Sucking louse

Biting louse

ancestors poured kerosene or light oils into cracks between floor boards to kill bed bugs. Vinegar was an accepted scourge for hair lice. After washing dishes, the soapy water was poured over rose bushes to kill the rose aphids. Even heat is an effective insecticide, and about 120° F. can kill most insects within minutes. In theory, household insects could be destroyed by closing windows in mid summer and stoking the fires, but heat takes considerable time to penetrate into bedding or wooden floors so this technique had obvious risks.

Over the centuries, some herbal materials or botanicals with strong insecticidal properties have been identified. Pyrethrum is legendary. This extract from the flower of the plant belonging to the Chrysanthemum genus has a powerful "knock down" capability that has been used for millenia. It was apparently known in ancient Persia and was cultivated in the Caucasus mountains and Yugoslavia. Today it is principally cultivated in Kenya and Latin America. In small quantities, it does not affect plants, birds, or mammals, including man, but only traces are needed to paralyze an insect and slightly higher levels will kill it. Because pyrethrum is very expensive, it is often combined with a slower-acting chemical to obtain the killing effect. Many synthetic pyrethrins have been developed over the past century but none are considered quite as safe and effective as the original.

Powdered tobacco was recommended as an insecticide in France as early as 1763, and nicotine from soaking tobacco leaves in water was a widely used insecticide for green flies and other species of aphids until recent years. But, while nicotine does not damage plants, it is a dangerous poison for mammals. Rotenone, the pulverized roots of some plants found in Southeast Asia and South America, was widely used as a fish poison until introduction into Europe as an insecticide during the mid-1800s. Rotenone does not harm plants, mammals, or birds, but human skin can become sensitized to it. Other botanicals include ryania and sabadilla; but, like the other botanicals, they vary in quality as an insecticide, are expensive, and are not easily available on a commercial scale.

As a result, numerous inorganic salts and gases were used as insecticides prior to World War II. The arsenic compounds included Paris Green (copper arsenite) that was used to control the Colorado potato beetle in the 1860s. But arsenic compounds also tended to kill plants, which did not gain strong favor with farmers. Flourine and cyanide compounds could be effective but were also highly dangerous. Furthermore, it was found that insects quickly developed such a resistance to these chemicals that amounts needed as an insecticide were sufficient to kill the plants being protected.

Colorado
potato beetle

Chlorinated Hydrocarbons

After World War II, synthetic insecticide sales rose sharply, and the most popular type was the chlorinated hydrocarbon, also known as the organochlorine or polychlor. The chlorinated hydrocarbon, which includes DDT and aldrin, had the advantage of low known toxicity to man, long persistence after application, and low cost. DDT, for example, is considered less toxic than aspirin for humans when swallowed, and a commonly quoted price has been fifteen cents per pound compared to about one dollar for the cheapest competing organophosphate. By contrast, the organophosphates, such as parathion, tend to be more toxic to man but considerably less persistent.

DDT was known to have been formulated in 1874 by a German Ph.D. candidate, Othmar Zeidlar, for his doctoral dissertation, but its potential was not realized and the work was forgotten until the late 1930s when Paul Müller, a Swiss entomologist, discovered DDT's insect-killing properties. The U.S. and Great Britain quickly developed the compound for use in World War II, which became the first major war when fewer soldiers died of communicable diseases spread by insects than by military action. Many thousands of civilians in Italy are believed to have been saved from a typhus epidemic when DDT replaced rotenone for control of body lice. After the war, DDT was credited with preventing millions of deaths by malaria, yellow fever, plague, and other diseases spread by fleas, house flies, mosquitoes, and lice.

Other chlorinated hydrocarbon insecticides were developed, including lindane, aldrin, dieldrin, and heptachlor. Some, such as heptachlor, were relatively toxic to humans though they were four to five times as effective as chlordane, another group of chlorinated hydrocarbons, for insects. Persistence appeared a valuable characteristic because the inside of a hut, for example, could be sprayed once for mosquitoes and the treatment would be effective for months and possibly years. This persistence, though, was also the chlorinated hydrocarbon's undoing. The chemical would degrade but the rate was slow and unpredictable, quickness depending upon moisture, temperature, surrounding media, and the decomposers available.

As the material spread and accumulated in organisms throughout the world's ecosystem, fishkills began to occur, birds died, and numerous subtle effects, such as the thinning of egg shells for some birds, began to be noted. Even more horrifying from the viewpoint of some users was the resistance that developed among insects as hardier strains of insects emerged and the pesticide became useless. Suspicion also began to develop that the low levels accumulating in mankind may not be as harmless as specialists were maintaining. In 1962, these worries found expression in an eloquent, somewhat overstated book, Silent Spring, by Rachel Carson. The impact upon government regulations and the insecticide industry was immediate. Chlorinated hydrocarbons began to recede from use, and the more persistent forms, including DDT, are now generally prohibited from use except in special circumstances.

Organophosphates

Organophosphate insecticides were by-products of Germany's development of nerve gases during World War II, and the chemical resemblance lingers. Typically, the insecticides are toxic but short-lived, remaining potent for a few hours or days. While efficient for killing insects, they are also poisonous to birds and mammals, including humans. Though the mechanism is not thoroughly understood and varies between organisms, the organophosphates appear to suppress an enzyme, cholinestrase, that regulates the impulses flowing through the nervous system. Without the enzyme, spastic uncoordination, convulsions, paralysis, and death can result. Some of the organophosphates are systemic, being absorbed and distributed throughout a plant so an insect can be killed by eating any part.

Because of the toxic characteristic, organophosphates were developed originally for agriculture and not for human contact. Nevertheless, they have posed a serious occupational hazard for agricultural workers and their families. At least 150 deaths, many children, occur annually from insecticide exposure, and most of these can be attributed to the organophosphates. Working conditions make the insecticide particularly hazardous. Protective clothing becomes uncomfortable in high temperatures. Workers are typically untrained and uneducated about proper handling of the chemicals. (Even though more toxic by ingestion or inhalation, most workers are likely to be poisoned by absorption of the chemical through the skin. Body areas differ in absorbtivity, the hands being relatively resistant while the crotch and back areas absorb nearly all of the insecticide reaching the skin.)

Parathion was the first widely used organophosphate insecticide, and it is still widely used on some crops, such as cotton. While not as toxic as some organophosphates, such as TEPP, parathion is still considered the leading pesticide killer of humans. In one widely publicized incident in 1967, seventeen persons died and over 600 became ill when flour contaminated with parathion was used for bread in a Mexican village. Children have died after playing with parathion containers, and workers have been poisoned after spilling a small container on their body or eating contaminated lunches.

Malathion is considered between 100th and 1,000th as toxic as parathion to humans and is often used in household vegetable or flower gardens. A normal human can rapidly metabolize malathion into relatively harmless materials, but parathion and some other organophosphates can sensitize a person so even a small dose of malathion can be fatal. About a dozen other organophosphates have been widely used; but, because of their reputations, they have generally been used with some caution. Also, insects have tended to become resistant to the organophosphates even more quickly than for the chlorinated hydrocarbons.

Carbamates

With the chlorinated hydrocarbons banned for general application because of persistance and the organophosphates considered too toxic and prone to resistance, popularity has passed to the carbamates. Examples are Sevin, a carbaryl

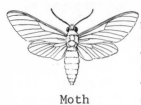

Moth

frequently used against gypsy
moths and other pests in
populated areas, and Baygon,
which is often used for cockroach control and insects
with public health significance.

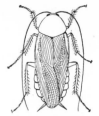

Cockroach

Like the organophosphates, the carbamates
apparently act against the enzyme, cholinestrase,
to affect muscle control, but they are more easily counter-
acted by the human body. Recent reports, especially among
workers in factories producing these insecticides, however,
have indicated an abnormal blood condition; and there is a
general expectation among specialists that more negative
health effects from carbamate exposure will be found. Furthermore, the history of other chemical insecticides suggests that an increasing number of insects will develop a
resistance to carbamates and their usefulness will decline.

Other Pest Control Chemicals

Several other broad insecticide categories, such as the
organic thiocyanates and the organic nitrogens, have been
developed. A similar situation exists for other forms of
pesticides. For rodents, there are the inorganic chemicals,
principally arsenic compounds, and botanical (plant) materials, especially strychnine. The usual rat poison, though,
is warfarin or a related chemical that inhibits blood clot-
ting so that the rats bleed internally. Rats resistant to
warfarin, however, have been reported from Great Britain and
several areas in the United States.

For fungi and bacteria that attack agricultural crops,
there are inorganic chemicals and several groups of organic.
Herbicides vary even more in their chemical composition.
Some are absorbed from the soil, while others are contact
chemicals. Some rapidly kill the plant, while others appear
to upset the hormone balance and cause growth abnormalities.
One group prevents a sprouted seed from growing. Because of
these variations, it is possible to use the chemicals on a
crop that has immunity or resistance to the killing mechanism while weeds are destroyed.

More chemical compounds are appearing annually on the
market. Research and development by the pesticide industry
has remained relatively constant, and about $71.6 million

were estimated spent in 1971. About 63,000 chemicals were screened in 1970 for pesticidal properties, but only eleven compounds survived the testing process to be registered for release to the market. About five years are usually required to test, develop, produce, and begin marketing of a new pesticide.

Meanwhile, the ability of insects and other pests to develop immunity or resistance maintains pressures on the industry. In the case of DDT, twelve insects were reported to be resistant by 1948, twenty-five by 1954, seventy-six by 1957, 137 by 1960, and 165 by 1967. When an insecticide is applied to a field, most of the insects--perhaps 90%--may be killed. But the remaining 10% has a reasonable probability of containing resistant or immune genes so they will be able to reproduce without competition and, because the predators may have been eliminated, without predatation. Reproduction rates for insects can be enormous. In theory a single pair of flies can produce over 100,000, 000,000,000,000 offspring in a single year. At the same time, most predator species, such as birds, are neither as numerous initially nor able to multiply as prodigiously. Furthermore, the resistant strain may have other genetic characteristics that are threatening. For example, the cabbage maggot strains resistant to cyclodiene lay twice the number of eggs and have double the life span of the nonresistant strain.

Aleochara attacking a cabbage maggot

Bioenvironmental Controls

Strengths and Weaknesses

Bioenvironmental controls are the population controls already existing in nature or ones that can be created by slight manipulation of natural conditions. The controls include:

(1) Utilizing biological control, including maximizing parasites, predators, and disease of pests.

(2) Increasing the resistance of desirable plants or animals to pests.

267

(3) Enhancing the genetic weaknesses of pests.

(4) Adjusting cultural factors, such as altering the timing, rotation, nutrition, watering, pruning, or land management for crops.

(5) Applying repellants or attractants.

(6) Creating physical or mechanical conditions incompatible with pests.

(7) Sterilizing pests by radiation or chemo-sterilants.

Natural Resistance

Some plants and animals have a natural resistance to certain pests. They are simply not attractive, presumably because of odor or taste. Others may be attractive but have another characteristic that foils the pests. For example, agricultural researchers in India are experimenting with a variety of rice that incorporates sand particles into the stem, dulling the mandibles of the stem borer and causing it to starve. In the first years of intensive genetic research with plants, the primary goals were higher productivity, improved product quality, and resistance to diseases. Now more effort is being focused on improving the ability of the plants to survive against pests. This is difficult, though, because the insects are also constantly developing new strains that enable them to overcome the natural resistance.

One frequently cited example has been the control of the Hessian fly with resistant varieties of wheat. Since 1942, about twenty-two varieties of resistant wheat have been developed. However, though new strains of the fly are constantly evolving and overcoming the resistance, this strength rarely extends beyond several varieties of wheat, and it is possible to substitute another resistant strain whenever outbreaks of the fly occur. Other crops undergoing intensive experimentation are cotton for the boll weevil; alfalfa for the spotted alfalfa aphid, leaf hopper, and alfalfa weevil; barley for the green bug; and corn for the European corn borer, corn ear worm, rice weevil, and corn root worm.

Boll weevil

268

Biological Control

Most serious pests in the United States, such as the gypsy moth, Japanese beetle, and corn borer, have been brought from another country where they were not regarded as pests because they were controlled by natural enemies. Compared to the effort expended, introduction of these natural enemies has been markedly successful, especially when compared to the immense effort needed to develop an acceptable chemical pesticide. Natural controls have been imported for an estimated 223 insect pests in the United States. At least partial success was achieved in 120 cases, and approximately seven examples of complete biological control have been recorded.

Japanese beetle

In a few cases, identification and evaluation of the control organism have been simple, and it has adapted completely to the new environment. In other cases, though, the process is time-consuming, laborious and not always successful. For example, at least thirty natural enemies for the California Red Scale have been introduced over the past seventy years. About eight have become established. Some have proven successful in the interior while others have thrived only on the coast. Some are not effective during hot summer or cold winter, while others can live under extreme conditions. Only by persistent searching, testing, and evaluation has relatively complete control been achieved.

The natural controls can be either predators, parasites, or diseases. One of the early successes of the biological approach was the control of the cottony cushion scale in California during the late 1800s. The California citrus industry was being destroyed, and the pest was becoming serious around the world except Australia, presumably its original home. Studies in Australia identified two control species, a parasite--Cryptochaetum iceryae--and the vedalia lady beetle. Within a year after these were introduced in California the scale had been controlled and the introductions were repeated successfully around the world.

The introduction of disease agents, especially viruses, has registered some successes. When the western grape leaf skeletonizer became established in California, a virus disease was introduced with sufficient success that all other efforts could be abandoned. In the Maritime Provinces of Canada, damage by the European spruce sawfly was curtailed by the accidental introduction of a virus disease. In 1963, the Environmental Protection Agency exempted a virus used on cotton from the requirement that only a minimum residue could be detected on crops. In other words, the virus was not considered to represent any threat to human health. Yet this contention evokes considerable skepticism by some observers. While the insect viruses being used have never been known to infect animals or plants, there is the worry that the insect viruses may become contaminated with other more dangerous types of viruses. And, as with many other control agents, the insects may be able to develop a resistance through their natural evolution.

Hormonal or Genetic Manipulation

By manipulating the hormonal balance or genetic traits of a pest, populations may be drastically reduced. For example, genetic strains have been identified among mosquitoes to produce mostly males. If this strain could be raised in the laboratory and released, it may be possible to diminish the ability of the species to reproduce. Among boll weevils, a strain has been developed that is unable to enter diapause, the suspension of activity necessary for survival during the winter.

Reaching the market stage is a group of artificial insect hormones that interrupt the natural development of insects. Eggs are prevented from hatching, larvae are deformed, and adults are sterilized. Major difficulties remain, however. Most of the artificial hormones are chemically unstable, expensive, and must be applied with exact timing. The effectiveness of some can be destroyed by only a day of exposure to sunlight. But insects will have difficulty developing resistance to juvenile hormones, and it appears probable that they will not be poisonous to man or animals.

Cultural Factors

Calculated changes in farming practices can also disrupt pests. Crops can be rotated according to a plan, fallow periods can be used judiciously, planting and harvesting can be precisely timed, and irrigation can be controlled. Control of soil nutrients, removal of plant waste, and use of pest-free seed are also being encouraged. For example, spring ploughing can destroy up to 98% of the corn ear worm paupae that survive the winter. Cotton seeds can be planted over a short period of time, limiting the number of boll weevil and boll worm generations possible. Since the boll weevil can use stalks for shelter during the winter, destruction of the stalks is important.

Repellants and Attractants

Insect behavior is governed by their physical state and external stimuli. Feeding and mating, in particular, can be regulated by natural specific chemicals, sounds, electromagnetic radiations, or a combination of these stimuli. Many female moths, for example, release a scent known as a sex pheromone, or sex attractant, to summon males. Some species of bees use scents to indicate location of food sources, and some ants chemically mark trails to their food.

Honey bee on comb

The chemical structures of many of these pheromones are being identified and synthesized to either trap or mislead insects. In a gypsy-moth infested area, strips of paper soaked in a sex pheromone have been dropped to overpower the natural female pheromone and, thus, reduce mating. Pheromones have long been used in gypsy-moth traps, the pheromone for each trap costing a fraction of a penny. On the island of Rota, an attractant for the male oriental fruit fly was laced with 3% insecticide and distributed by airplane. Within several months, the species had been eradicated from the island.

Oriental fruit fly

Chemical repellants have been considered more successful for rodents and other mammals than for insects. With varying degrees of effectiveness chemical repellants have been used against rats, mice, squirrels, deer, and rabbits. Chemical coatings on roosting areas have been used against birds, especially pigeons, with some success. Repellants to prevent birds from eating seeds, however, have not proven adequate against the vigor of a bird's appetite.

Sterilization

Releasing sexually sterile insects to mate with normal insects has been an effective strategy with some species. If the female mates only once and this mating is with a sterile male, a large proportion of sterile males will rapidly cause the total population to collapse below a population density where fertile males can locate a female. Perhaps the most spectacular success was in eradication during 1958-59 of the screw worm, a serious parasite of livestock in the southeastern United States. About two billion flies were reared, sterilized, and released over 70,000 square miles.

Screw worm adult

A similar approach has been used to prevent the entry of Mexican fruit flies into southern California. This technique, though, involves immense logistical problems in rearing and sterilizing billions of insects needed to flood a target insect population. At the same time, these released insects must retain healthy and aggressive characteristics so they can compete for mates. A variation on this approach has been to develop a hormone that produces contagious sterility for one generation. Laboratory-reared insects are fed the hormone and released. Like an insect version of venereal disease, the sterility can be quickly spread through an insect population if they mate frequently.

Other Approaches

Numerous other methods of controlling pests are being used or could be developed. By restricting the movement of plants and animals across U.S. borders, some pests are presumably hampered in entering. Areas having serious outbreaks of pests are sometimes quarantined. For insects, electromagnetic energy of various types can be a control mechanism. Some energy waves raise body temperatures, killing the insects. Infrared, ultraviolet, and invisible light can be used in insect traps. Some insects are discouraged from breeding or feeding by introduction of light. And, for some pests, the most reliable and selective device yet invented may be that venerable invention, the fly swatter.

Integrated Control

Official Policy

The U.S. Department of Agriculture has stated that its official pest control policy is "integrated pest management." Primary reliance is upon bioenvironmental controls but chemical controls are used when justification exists. A typical management approach is used. An analysis of potential pest problems is made, and preventive measures, such as use of resistant plant strains and modified farming practices, are followed. Then fields are monitored to determine the levels of pests, their natural enemies, and important environmental factors. Only when significant, preventable crop damage from a specific pest appears inevitable will suppressive measures be taken. Even then, chemical controls are used only after bioenvironmental approaches have been ruled out.

The emphasis is upon diversity of approaches. Insects are viewed as shifty, adaptable, resilient, and unpredictable creatures. To plan a control strategy, such as scheduled spraying, without considering all conditions at the time is regarded as an invitation to disaster.

Complications

Despite the pressures of logic and governmental blessing, integrated pest management has not been widely applied. The reasons are varied but each appears to reinforce the other. The flexibility and complexity of the approach contradicts the farmer's psychological attitude and educational background. Also, the reward system in our industrial society does not favor bioenvironmental strategies. Basically, perhaps, the public attitude towards science and technology has not generated the sympathy necessary for the essentially scientific approach to pest control.

Farming is a risky and difficult occupation. A farmer must battle the elements, diseases, insects, and market trends for his products. If a natural disaster strikes his crops, he loses and his family suffers. The farmer's self-image, therefore, is one of the rugged individual battling the harsh forces of nature. Insects are part of nature, and they must be vanquished with poison or whatever the most

powerful tool may be. Chemical insecticides satisfy this
yearning.

In addition, the farmer already has enough worries
without adding the research and flexibility implicit in an
integrated management approach. As much as possible, everything should be planned in advance. Weather permitting,
ploughing occurs on schedule. Fertilizer is spread at a
time and in a quantity previously estimated. A fixed amount
of seeds are planted. Thus, it follows naturally to counter
insects on a scheduled, predetermined basis by the spraying
of chemicals.

This mode of thinking is encouraged by the purveyors of
chemical pesticides. In farming areas, the pesticide sales-
man provides "free" services as a diagnostician, and he then
prescribes the cure, which is typically a chemical pesticide. He is not licensed. He does not have any interest
in promoting any approach except the application of pesticides because merchandising is his essential responsibility.
Even governmental extension agents report that they are constantly courted by the chemical companies, which provide a
steady flow of information subtly directed towards chemical
products. The problem may have even deeper roots. No one
can patent a wasp or naturally found virus, but chemicals
can be patented and marketed.

To some degree, the problem may be related to the public mood towards application of science and technology for
solving public problems. People are suspicious towards any
involved, highly technical strategy. A bioenvironmental
method or chemical control method is simple and understand-
able. The reasoning and planning necessary to apply these
within an integrated management concept is not as easily
understood or accepted.

Governmental Regulation

Before 1972

The dangerous potential of chemical pesticides was
first acknowledged in the Insecticide Act of 1910, which
applied simple regulation to the few pesticide products then
available, such as Paris green and pyrethrum. With the
flood of new chemicals appearing on the market after World
War II, the need for expanded regulation was apparent, and

274

the Federal Insecticide, Fungicide, and Rodenticide Act of
1947 was enacted. This established a system of registration
and labelling for all pesticides being marketed. Manufacturers were required to present evidence that the products
were safe when used as directed and possessed the effectiveness claimed on the label.

In 1959, the Act was amended to cover herbicides, repellants, and other new classes of pesticides. Still two
other weaknesses in the Act were becoming apparent. One was
a loophole known as the "registration under protest" section
that permitted marketing of unregistered pesticides while a
full investigation of its effectiveness was being completed.
Since numerous devices existed to postpone completion of an
investigation, this section in the law was eliminated by an
amendment in 1964. The other weakness was more basic and
difficult to counter with an amendment. The existing law
had as its primary objective the direct protection of humans
from poisoning. It did not have a mechanism for protecting
them against a chemical that damaged the natural environment
and, indirectly, humans. Nor could a product be easily
withdrawn after this type of damage had been identified, as
occurred in the case of DDT.

Federal Environmental Pesticide
Control Act of 1972

In 1972, the pesticide legislation was rewritten to
cope with the broader environmental problem. This time, the
Act provided for registration of products under two classifications, general use and restricted use--or both if some
uses could be considered in isolation from others. For any
restricted use, the pesticide could only be used by an ap-
plicator certified by the Environmental Protection Agency
or, if an acceptable state plan were prepared, by the
states. The Act further provided for experimental use of
a pesticide so a manufacturer could collect the information
needed for registration.

Besides the authority to suspend the registration of a
pesticide immediately if its continued use would represent
an "imminent hazard," the administrator of the Environmental
Protection Agency must cancel the registration of every
pesticide on the market at the end of each five-year period
unless the manufacturer applies for continuation of registration. In other words, the burden of justifying continued

marketing of a pesticide product falls upon the manufacturer and is not regarded as an automatic right.

Other Laws

Because of the dangers represented by economic poisons, numerous other laws and regulations also affect pesticide use. In a 1954 amendment to the Federal Food, Drug, and Cosmetic Act of 1938, any pesticide chemicals "added to a raw agricultural commodity shall be deemed unsafe . . . unless--(1) a tolerance for such pesticide chemical . . . has been prescribed, or (2) the pesticide chemical has been exempted from the requirement of a tolerance."

Furthermore, the Secretary of Agriculture must certify that the pesticide chemical is useful for the purpose identified for the tolerance or exemption permit.

Because most pesticides are classified as "Class B" poisons, their transportation falls under the jurisdiction of the Hazardous Materials Regulation Board of the Department of Transportation and must comply with specific rules on containers, labels, and handling. Since widespread dispersal of pesticides and other chemicals can occur by spraying from aircraft, the Federal Aviation Agency has issued regulations requiring pilots flying spraying equipment be tested on their knowledge of these materials and their proper use. The Federal Trade Commission also has responsibilities for any advertising that "contradicts, negates, detracts from, or is inconsistent with any statement, warning, or direction for use, in the labelling of such product . . . ."

Encouraged by the 1972 federal law, individual states have been enacting laws to qualify for certification of applicators. Even more local communities are enacting ordinances, spurred by the state and federal activity and the realization that high-powered spraying equipment is increasingly being used by commercial gardeners and other operators not trained or necessarily responsible in their attitude for use of chemical poisons. A model law has been prepared and distributed by the Environmental Protection Agency.

Conclusion

Pest control contains the elements that characterize most of our serious environmental problems. Continued use of pest control chemicals appears necessary for survival of our altered, urbanizing habitat, but we realize that use of these chemicals must be regulated or they, in turn, will be deadly destructive. The basic problem that emerges is not an essential inadequacy of our technology but in the conflicting demands and attitudes among the people and the institutions they represent. The housewife today expects fruit and vegetables to have an appearance approaching perfection. The farmer often retains beliefs from a simpler era and wants his work uncomplicated by complex decisions. Industrialists want to promote chemical products and resent government advocates of bioenvironmental approaches. Disillusioned citizens want to escape to a simpler life where misused technologies will not harass their nerves and sensitivities. Legislators hesitate to enter this emotion-charged arena of political conflict. It is not surprising that we have been slower to enact revised legislation in this area than for some other environmental problems.

QUESTIONS FOR DISCUSSION

1. Name six insects, birds, or animals commonly regarded in your community as pests and then identify circumstances where each might be considered a desirable species.

2. Describe three bioenvironmental control techniques you could apply in your apartment or house if it were suddenly infested with cockroaches.

3. Assume that trees in your suburban community were devastated last summer by gypsy moths; and, since a repetition next summer is widely expected, a group of citizens have applied to the Department of Health for a permit to aerial spray the wooded residential area early next spring with Sevin, a carbamate. What would be your reply if the mayor requested your advice for the most appropriate action?

4. Compare contamination of the natural environment with chemical pesticides to regional air pollution. How are (a) the physical manifestations and (b) underlying social and economic causes similar?

SUGGESTED READINGS

Three general categories of readings on pesticide use are available. First, numerous books have been written for the general public. Style and accuracy vary immensely. Everyone in the environmental field, though, should be acquainted with Rachel Carson's Silent Spring, which helped precipitate the environmental awareness of the sixties. Many similar books have followed. A related type are the numerous, delightful books written by naturalists to open the insect and animal world for better understanding by the human species. One is listed below.

1. Rachel Carson, Silent Spring (New York: Houghton Mifflin Co., 1962).

2. Howard E. Evans, Life on a Little Known Planet (New York: E. P. Dutton & Co., 1968).

Second, there have been numerous governmental reports and proceedings on scientific symposia that may be useful in better understanding of the subject. Only a few are listed below.

3. Restoring the Quality of our Environment, Report of the Environmental Pollution Panel, President's Science Advisory Committee (November 1965), U.S. Government Printing Office, $1.25.

4. Report of the Secretary's Commission on Pesticides and Their Relationship to Environmental Health, a Report to the Secretary of the U.S. Department of Health, Education and Welfare (December 1969), U.S. Government Printing Office, $3.00.

5. National Academy of Sciences, Scientific Aspects of Pest Control (February 1-3, 1966), Publication 1402, $5.00.

6. National Academy of Sciences, Pest Control: Strategies for the Future (1972).

7. Council on Environmental Quality, Integrated Pest Management (November 1972), U.S. Government Printing Office, 55¢.

The agricultural departments in both the federal and state governments provide a wide range of free reference materials. Write to your state government for a list of free or inexpensive booklets on local pests and their control. The U.S. Government Printing Office can provide a similar list for the U.S. Department of Agriculture. An excellent and little-known bulletin on current knowledge of chemical pesticides is furnished by the Center for Disease Control. These can be obtained from the Technical Development Laboratories, Center for Disease Control, U.S. Department of Health, Education and Welfare, Savannah, Georgia 31402. Ask for the bulletin on Public Health Pesticides.

Chapter 14

BROADER ASPECTS OF ENVIRONMENTAL MANAGEMENT

>"Man's mind cannot grasp the
>causes of events in their com-
>pleteness, but the desire to
>find those causes is implanted
>in man's soul." Leo Tolstoy
>(1869)

Role of Public Participation and Education

The "People Factor"

The essence of environmental management is the influencing of human behavior to maintain environmental quality.
Acts that diminish quality are discouraged while those that
enhance quality are promoted. Positive attitudes towards
environmental quality have to be created, and the people
must be motivated to act upon those attitudes.

These objectives involve attaining three successive
stages: (1) perception, (2) action, and (3) coordination
in problem solving. Perceptions must exist since people
cannot act positively on concepts that are absent from their
minds. Perception is a skill that must be developed and
exercised. Few people can comprehend the full scope of
their daily environments. There may be a vague awareness
that space and objects have aesthetic or life-giving qualities, but the relationship will normally be distant and void
of personal meaning unless deliberately fostered. Eyes can
pass over colors, textures, lighting, and the distribution
of space and objects just as the mind passes without comprehension over a dead and dying environment.

Every great culture has environmental values. In the
folklore of architectural planning, the North Europeans are
oriented towards natural phenomena and the presence of flow-
ers, trees, shrubs, and grasses. The Mediterranean emphasis
is upon urban forms, such as the juxtaposition of structures
and spaces as, for instance, about a plaza. A similar ster-
eotype is the Oriental stress upon harmony with nature and
the growth of appreciation through concentrated contemplation

of simple natural objects--a stone, a flower, a weed, or a seascape. Yet the common trait of environmental perception links all.

Perception inevitably influences human behavior; but, for maintaining a quality environment, the behavior has to be directed towards specific acts, such as the placing of paper in trashcans instead of throwing it on the ground. Furthermore, the specific acts must take precedence over other possible acts that reflect a different hierarchy of values, such as the saving of personal energy needed to walk to the trashcan. Personal habits reflect an individual's value priorities, and considerate treatment of the environment requires emphasis on environmental values. Public information and education is required, especially in developing the attitude known as the environmental ethic, the feeling that trees, animals, and even the architectural works of previous humans deserve respect and have inherent "right," if only to serve a future environment.

At a third stage, environmental management must influence group behavior by enlisting participation of many individuals towards attaining common environmental goals. A degree of consensus must be attained on the definition of environmental quality. Priorities within the society have to be established for environmental goals. Means of compensating those who are innocent losers, such as the householder who must vacate his land for a new park, have to be determined, and equity considerations would reduce the gain of nonproductive winners, such as owners of property adjoining a new highway or another public investment.

Gaining Public Participation

Enlisting public participation is generally gained by moving through the three stages of consciousness-raising, creating action-influencing attitudes, and gaining problem-solving cooperation. Initially, some estimate should be formed of the existing state of public opinion and inclinations. Attitudinal surveys can provide understanding of some aspects, and interviews with opinion leaders can develop insights into the background and dimensions of attitudes. Consciousness raising consists of informing the audience of the relevancy of a phenomenon to their lives. No single technique will reach an entire community, unless it is unusually cohesive; but, by judicious identification of communication networks, a carefully planned campaign can reach the bulk of the politically active persons. Informative articles in newspapers, documentaries on television, and appropriate news releases for the radio are used. Flyers and posters may help if a receptive audience can be expected to see them. Personal communication networks, such as presentations at club meetings and items in newsletters, can reach still other groups.

Active participation is gained by offering an opportunity for expressing interest on real issues, especially when the context indicates that participation may effectively influence an outcome. Public hearings, for example, can mobilize wide public support if the location and time of the hearings permits those with stakes in the issue to attend. An election, especially with a referendum, will place an issue into the political arena where the traditional and organized skills of political persuasion can operate.

Formulation of wise solutions to environmental problems cannot be left to the chance interaction among few participants. The analytical and conceptualizing capabilities of a trained staff are usually needed to pose the issues and desirable outcomes in the clearest and most persuasive forms. Political leadership of a constructive type is irreplaceable. A skilled and foresighted politician will anticipate future needs of his constituency and will try to educate or persuade them to accept the appropriate course even though it may not be immediately necessary. Successful laws and decisions are those that both answer immediate problems and prevent potential ones from developing by influencing social change constructively. These are political acts; and, in our society, they are ultimately the responsibility of elected representatives.

Environmental Education

Ideally, programs to raise public consciousness and enlist participation would not be necessary because the awareness and willingness to maintain a quality environment would already exist. This can be achieved only by environmental education, which has two aspects, youth and adult. Numerous educational programs for schools have been organized, and others are continuing to be formulated. With some federal aid, many educational materials--books, posters, films, and instruction manuals--have been prepared and distributed. Nature trails have been developed and instructors trained. Children today probably have a stronger interest in protecting environmental quality than previous generations, and the impact of this awareness can be expected in public policy during approaching decades.

Public interest groups and the mass media have provided the key function of adult education. Some groups, such as the League of Women Voters and the Sierra Club, have skillfully prepared easily read and technically accurate explanations of environmental problems. A necessary step in adult education within our society, however, appears to be the political process with the adversary exposure of facts and viewpoints. This exposure also serves as a catharsis for opposing interests.

Environmental Management as an Art

The Need for Judgment

Previous pages have emphasized a technique for approaching environmental problems. The situation is studied to identify the problem; alternative actions are defined; a plan is prepared and implemented; and actions are adjusted as effects are registered. The approach appears highly systematic, and the assumption may be instilled that all factors can be considered mechanically and comprehensively. This is impossible, however; and the problem-solving technique can only serve as a guideline, not a formula. Discretion and judgment have to be applied at every step. The situation can be compared to a set of carpentry tools that can assist in building a cabinet, but the quality of the cabinet depends upon the skills and judgment of the carpenter using the tools.

Inherent Weaknesses in the Problem-Solving Approach

At least three overlapping weaknesses in the problem-solving approach to environmental problems test management skills: (1) interpreting the spirit of the approach, (2) predicting future conditions, and (3) distinguishing pertinent information. These weaknesses are often reflected in environmental planning documents, such as environmental impact statements. A weak document will dwell excessively on intricate organizational details that do not enhance the report's utility for decision making. Elaborate but almost meaningless charts and tables proliferate. The future is predicted by simple extrapolation of past events with no allowance for uncertainty. Reportable but unimportant facts are recited for pages while critical but vague information is considered in a sentence.

Interpreting the Technological Approach. The technological (problem-solving) approach arouses deep apprehensions among many thoughtful members of our society, and their misgivings have been articulated by numerous social and political philosophers. Most concerns can be summarized in three categories.

(1) If interpreted rigidly, the standardization tends to suppress spontaneous and creative expressions. The pressures to conform tend to discourage thoughts that cannot be cast into a "scientific" empirical mode. No environmental impact statement, for example, can distill the sweetness of spring air or the haunting beauty of a forest in shimmering moonlight with the same concise detail as the economic value of an ore body or the marketing potential of timber. Nor, critics charge, should it be necessary to debase the aesthetic sweetness of life by trying to compare it to the drudgeries.

(2) In its fascination with technology, our society tends to become increasingly materialistic but insecure. Means become confused with ends because we have failed to define ends. The processing of reports becomes an end, not a means. Assumptions are forgotten and purposes are distorted to avoid disrupting the organization of technicians.

(3) The political structure potentially becomes more dictatorial as the population accepts standardization of political processes and behavior under the guise of technical efficiency, a means. Political dissent is, in bureaucratic terms, inefficient. Pressures mount on the population to submit to the prescribed decisions. Also, the increasing concentration of economic capital and technical knowledge--both sources of power--become increasingly centered in a few individuals.

In other words, observers fear that our activities will become progressively dehumanized, sterile, and submissive. We will be busier but will achieve less satisfaction. At the same time, rational men would agree that our urban society is dependent upon technology and the technological approach to problems for both our necessities, including

284

food, housing, clothing, and public order, as well as our aesthetic culture.

Prediction. "Management" implies an orderly progression of events made possible by influencing causative factors. Events are shaped before they happen. This necessitates some understanding of factors causing a future event and the probable outcome without and with interference. In other words, accurate predictions--forecasting with a degree of commitment--form a key element in management. Predicting in an environmental field, though, is risky. Natural ecosystems are extraordinarily complex and are in constant change. Equally difficult is the prediction of human behavior when a wide range of choice exists. Human attitudes are fraught with contradictions, and some individuals will always express opposition to emphasize their individualism. Yet commitments are implicit in decision making, so evidence does have to be arrayed, analyzed, and interpreted to create a prediction, although imperfections are realized.

Discriminate Selection of Information. Judicious use of information in a report always provides a harsh test of skills. Immense amounts of numerical data, such as numbers of varying reliability describing climatic or geographical conditions, are always available; but these rarely describe the critical factors in the problem. The rules are simply stated. Available information should be sifted and only parts directly relevant to the problem should be synthesized and presented in a concise fashion that will contribute to a wise decision. Information that is difficult to obtain should be sought; and, if unavailable, formulated from known clues. But this ability comprises part of the skill and artistry of environmental management.

Opportunities for Environmental Management

The Personal Dimension

Environmental management has a personal dimension in everyone's life. We live in an environment, and contributing to its maintenance is a continuing responsibility. Since distinguishing quality requires taste and judgment, individuals should express their individual preferences. Because the tangible rewards of quality environments are often slight for governmental officials, some vigilance by citizens is justified. But judicial social pressures against other citizens are also needed when they are violating public rights in the environment. And, increasingly, most people are having to curb their own personal lifestyles to conform with the realities of resource limitations and the collective impact upon common water, air, and land areas.

Professional Opportunities

On a professional level, the opportunities are extraordinarily diverse. Usually, though, a choice can be between three broad categories of employers: government agencies, environmental interest groups, and private consulting or industrial firms. In all cases, professional openings vary in responsibilities, and it is impossible to describe the wide variety of tasks that are needed. From a broad, managerial viewpoint, the federal government probably offers the most diversity. The agency with the most direct involvement is, of course, the Environmental Protection Agency. Most of their openings have highly technical requirements, but they also seek some environmental generalists. The Council for Environmental Quality also employs individuals with extensive planning or managerial backgrounds, but the Agency is relatively small and opportunities are limited.

Beyond these agencies, almost every federal department that has responsibilities for construction or decision-making in the natural environment requires some professional talent for the tasks. Environmental impact statements are a major responsibility within the Department of Transportation. The Department of Agriculture has varied responsibilities in the environmental field, and individuals seeking openings should also contact the Department of Housing and Urban Development, the Department of the Interior, and the Department of Defense, especially the Corps of Engineers. After some experience in the field, the Office of Management and Budget may also be considered.

This range of opportunities is usually duplicated at the state level. Again, an Environmental Protection Department usually exists, but some of the responsibilities may be shared by the Department of Health. Offices of economic or regional planning should also be considered, and transportation departments typically have to make heavy investments in environmental planning. County and city governments provide still a third level that should be explored.

Public-interest organizations, such as the Sierra Club or the Environmental Defense Fund, offer opportunities for individuals with special talents, especially writing or law, and unusual motivation. Salaries are generally low, political involvement is typically high, and a person can easily become disillusioned if not mentally prepared.

A large proportion of the environmental designs for projects and the preparation of environmental impact statements occurs at the private level, usually in a consulting firm. Large engineering offices dominate the field, but there are also many private research or small consulting agencies that contribute significantly and offer interesting, pleasant challenges. Within industrial corporations, almost every major manufacturer or developer of resources today has positions with primary responsibility for environmental affairs. Sometimes these are coupled with occupational health, but these are increasingly separate. The nature of the opportunities and the degree of authority for investing funds in improving conditions will vary from firm to firm.

Conclusion

One interpretation of this book's purpose could be that it describes how we can control some of the forces mankind has unleashed in our environment. There are always temptations to claim that one's present era is uniquely critical and complex. Ample reasons can be found in today's world to make this claim. Rapid depletion of resources needed to sustain present styles of life is occurring. Population pressures are condemning open spaces and altering land uses that had predominated for millennia. Economics dictate industrial growth that constantly threatens pollution and raises questions about man's wisdom with his new technological toys. Biotic diversity of the globe is decreasing, and nuclear holocaust could momentarily erase the environment as we know it.

So are these times dismal? I think not. In economics, there is the principle of the Hidden Hand: For every imminent disaster, there is a comparable opportunity. These are times crammed with awesome and exciting events. Artistic skills and cultural expressions have reached superlative sophistication. New materials, such as plastics, allow a simpler, more comfortable, and diversified life. Social and economic mobility has never occurred on as massive a scale as in contemporary nations, such as India and China. A century ago, most of us would have been constrained to a life of tilling exhausted soil in abject poverty. Now we can travel and view the world's treasures in a style unknown to bygone kings. Most importantly, we can influence both our destiny and the quality of the environment about us.

In 1859, Charles Dickens wrote the oft-quoted introductory phrases to The Tale of Two Cities. The remainder of the long sentence is less known, but it more aptly expresses the turmoil of our age. "It was the best of times, it was the worst of times, it was the age of wisdom, it was the age of foolishness, it was the epoch of belief, it was the epoch of incredulity, it was the season of Light, it was the season of Darkness, it was the spring of hope, it was the winter of despair, we had everything before us, we had nothing before us, we were all going direct to Heaven, we were all going direct the other way--in short, the period was so far like the present period, that some of its noisiest authorities insisted on its being received, for good or for evil, in the superlative degree of comparison only."

QUESTIONS FOR DISCUSSION

1. What forms of environmental education have you seen in the past twenty-four hours? What stages of environmental participation did they represent? Can you always distinguish stages?

2. Design an ideal environmental education program for (a) fourth grade school children and (b) high school students using resources in your community.

3. Prepare a case study describing an environmental issue that was decided politically in your home community. What were the apparent educational effects?

4. Identify two examples in national governmental administration when one or more of management weaknesses were apparent. What are some classical historical examples?

5. Prepare a list of ten agencies or occupational categories, other than those identified in this chapter, where a specialist in environmental management may find employment.

SUGGESTED READINGS

For information on environmental education in schools, refer to the Environmental Education Report, 1621 Connecticut Avenue, N.W., Washington, D.C. 20009. Also, the Division of Technology and Environmental Education, Office of Education, Department of Health, Education and Welfare, in Washington provides grants in environmental education and can be a source of information.

Among the numerous noted authors, including Thorstein Veblen, Lewis Mumford, Sigfried Giedion, and Robert Merton, who have examined the interactions between technology and society, perhaps the single most stimulating volume would be Jacques Ellul's The Technological Society (New York: Alfred A. Knopf, 1964) with the introduction by Robert K. Merton.

For individuals considering a career in the environmental field, Opportunities in Environmental Careers (New York: Vocational Guidance Manuals, 1971) by Odum Fanning may be helpful. Also, one could probably gain insights and ideas by browsing through one of the recent directories on environmental organizations and information sources, such as the National Foundation for Environmental Control, Inc.'s Directory of Environmental Information Sources (Boston: Cahners Books, 1972).

Urban Development Corp. (N.Y.), 62
Urban environment, nineteenth century, 80

Viruses, in water, 75

Warfarin, 266
Waste heat, 43, 78, 84, 104, 137ff, 152, 175
  agricultural use, 153
  amount, 148
  disposal, 155
  nuclear vs. fossil fuel, 148
Waste heat utilization (see Heat, utilization)
Waste minimization, 237
Water:
  aesthetic effect, 69, 70
  agricultural use, 70
  chemistry, 34, 67, 73
  cooling, 155-156
  heat added, 139, 141
  industrial use, 70
  need for, 69
  properties, 73ff
  role in ecology, 70-71
  temperature, 68
  testing, 75-76
  turbidity, 79
Water conditioning, 154
Water consumption (see Water use, amount)
Water cycle, 36, 68
Water for cooling, 137-138, 151-152
Water pollution, 43, 71, 72, 84, 107, 115, 137
  chemicals, 108
  from landfills, 253
  in estuaries, 122
  viruses, 108
Water Pollution Control Act (1948), 102

Water Pollution Control Act (1956), 102
Water Pollution Control Act (1972), 15, 108
Water quality, 71, 81 (see also Water pollution)
  legislation, 102
  standards, 102ff
Water Quality Act (1965), 102
Water Quality Improvement Act (1970), 103
Water requirements, 69
Water supply, 80
  chemicals added, 88
  pipelines, 108ff
  settling, 88
  shortage, 107ff
  sources, 87
  standards, 88
  storage, 88
  treatment, 88ff
Water supply system:
  leakage, 90, 108-109
  urban, 84
Water table, 87
Water temperature, 154
  effect on fish, 143
Water treatment, 85
  plant faults, 101
  processes, 88ff
  variations, 100
Water use, 69-71, 138
  amount, 85-86
  variations in demand, 86
Wetlands (see Estuaries)
Wind, air pollution, 180
Wind energy, 151
Worms, in water, 75

Zoning, 61